Contents

Wet Storage

Storage Technology

IMechE
Conference Transactions

I MECH E
150th Anniversary
1 8 4 7 - 1 9 9 7

International Conference on

Storage in Nuclear Fuel Cycle

18–19 September 1996

Organized by the Nuclear Energy Committee of the
Power Industries Division of the
Institution of Mechanical Engineers (IMechE)

IMechE Conference Transaction 1996 – 7

Published by Mechanical Engineering Publications Limited for
The Institution of Mechanical Engineers, London.

© The Institution of Mechanical Engineers 1996

ISSN 1356-1448
ISBN 0 85298 998 9

A CIP catalogue record for this book is available from the British Library
Printed by The Ipswich Book Company, Suffolk, UK

D
621·4634
Im

Applications and Licensing

Related Titles of Interest

Title	Author	ISBN
Nuclear Decommissioning	IMechE Conference 1995–7	0 85298 955 5
Emergency Planning and Management	IMechE Conference 1995–6	0 85298 954 7
Sizewell 'B' – Aiming to be First	IMechE Conference 1994–1	0 85298 903 2
Nuclear Power Plant Safety Standards: Towards International Harmonization	IMechE Conference 1993–8	0 85298 860 5
Achieving Efficiency through Training – the Nuclear and Safety Regulated Industries	IMechE Seminar 1994–7	0 85298 944 X
The Inspection and Structural Validation of Nuclear Power Generation Plant	IMechE Seminar 1993–14	0 85298 878 8

For the full range of titles published by MEP contact:

Sales Department
Mechanical Engineering Publications Limited
Northgate Avenue
Bury St Edmunds
Suffolk
IP32 6BW
England

Tel: 01284 763277
Fax: 01284 704006

C512/Key/96

Keynote Address

B J FURNESS BSc, MSc, DIC, MIMechE, **D J MASON** BSc, MSc, MSRP, **K K McDONALD** BEng, CEng, MIChemE, and **P DICKENSON** MBA, MSc, CEng, MIMechE
HSE Nuclear Installations Inspectorate, UK

1. PERSONAL INTRODUCTION

You will have noted from the Chairman's introduction that my interest is primarily in nuclear safety. You note that I say primarily, because HSE, in considering what is reasonably practicable, can never divorce itself from the realities of technical and financial constraints. This conference provides us with an opportunity to discuss the technology and options for the management of radioactive materials arising from the nuclear fuel cycle. This includes not only spent fuel but also the radioactive wastes, and plutonium and uranium which are produced as a consequence. My aim is to set the scene and stimulate your thinking so that, as an industry which is world wide, we ensure safety now, avoid complacency and maintain a commitment to achieve the highest standards of safety into the future.

2. BACKGROUND

2.1 The nuclear industry has been in existence for half a century. Whilst some countries currently consider spent fuel to be waste, others are committed to the policy of the fuel cycle, in which uranium and plutonium are extracted from the spent fuel by reprocessing. Although this process generates low, intermediate and high level radioactive wastes, the uranium and plutonium are available for potential reuse in the manufacture of further fuel. There are currently no recognised disposal facilities for spent fuel or for high and long lived intermediate level radioactive wastes. Therefore, large quantities of radioactive materials have been accumulated throughout the world.

2.2 Management options for radioactive materials are unlikely to be chosen on technical grounds alone; selection is inextricably linked with national policy and international standards, as well as security considerations, economics and, what is becoming increasingly important, public perception.

2.3 Although the technology for direct disposal of spent fuel has not yet been proven, you will hear in the papers to be presented some of the options available for the management of spent fuel including prolonged wet or dry storage, and direct disposal following an interim storage period. The reprocessing option involves storage of the spent fuel before processing and the storage of the resulting wastes and by-products afterwards. Thus, storage, whether it be for months or years, is a fundamental requirement whichever option is pursued.

2.3.1　According to International Atomic Energy Agency estimates, more than 165,000 MTHM (metric tonnes of heavy metal) of spent fuel from nuclear power reactors have been discharged worldwide. 110,000 MTHM of this fuel are currently in storage with the remaining 55,000 MTHM having been reprocessed, including about 40,000 MTHM at BNFL's Sellafield site. By the year 2010, it is estimated that 300,000 MTHM will have been discharged from nuclear power stations worldwide, of which 200,000 MTHM is expected to be in storage, an approximate two fold increase in current stocks, with the remaining 100,000 MTHM reprocessed. Commensurate with this is an increase in stockpiles of uranium and wastes. Plutonium stocks, however, will depend greatly on the take-up of Mixed Oxide fuel. As existing on-site storage facilities for these materials become full, management decisions are needed which clearly have an impact into the long term. The options include:

    (a) increasing capacity in existing stores;

    (b) building new stores, which may utilise wet or dry store technology, constructed on a single site or on a number of sites; or

    (c) "buying" storage services.

2.3.2　Where fuel reprocessing has been carried out, for example by Cogema at Cap La Hague, France, or by British Nuclear Fuels (BNFL) at Sellafield, UK, significant stockpiles of recovered uranium and plutonium have been generated. Not only do these have to be stored safely, but strategies for their future management require consideration to ensure continuing safety and to ensure that proliferation does not become a problem.

2.4　In the UK, some radioactive wastes arising from fuel manufacture, reprocessing, research and military programmes have been stored in on-site facilities for 30 or 40 years. A number of waste facilities are approaching or have passed their design lifetimes and their suitability for further long term storage is being questioned; there are examples of waste storage in silos where, for historical reasons, there has been little or no segregation, conditioning or packaging. There is uncertainty about when the national Nirex repository will be available (the current estimate is 2012) and its operating strategy is undecided. This inevitably impacts on projected storage facility requirements and the potential hazard to workers and the public. Pressure is now on waste producers to retrieve, process and store wastes in a state of passive safety, that is in a form requiring minimum human intervention and which will be acceptable to Nirex without further repackaging.

x

3.   INTERNATIONAL SCENE

3.1   Turning now to the international arena.

3.1.1   An important area of HSE/NII involvement is in the current negotiation with Member States of the International Atomic Energy Agency, on an international Convention on the safety of radioactive waste. HSE is represented in the UK's delegation team together with the Department of Environment, which is taking the lead, and the Environment Agency. It is likely that the UK will become a contracting party to the Convention when it comes into force. The Convention will be of an incentive type and will not therefore require waste management facilities to be inspected by other countries. Instead, each contracting party will prepare a report indicating how each of the articles in the Convention is met when regulating facilities under its control.

3.1.2   The UK will need to demonstrate how its legislative, regulatory and administrative procedures meet the obligations under the Convention. This will also involve a demonstration that for all stages of radioactive waste management the individual, society and the environment are adequately protected.   The UK report will probably describe the relevant regulatory bodies and show their independence from industry. It will demonstrate how the industry has prime responsibility for the safety of radioactive waste management and indicate the steps taken by the regulators to ensure that this happens.

3.1.3   The Convention is likely to cover most of the UK's radioactive wastes and their respective storage facilities. Spent nuclear fuel will be included in the Convention. Reprocessing of spent nuclear fuel is not likely to be covered, although the resulting radioactive waste will probably be within the scope of the Convention.

3.2   Long term management strategies for spent fuel differ from country to country. The question facing every user of nuclear power is direct disposal or reprocessing. The subject has raised much global deliberation. Of the 31 countries with nuclear power programmes, 5 have currently opted for direct disposal for some or all of their spent fuel, 8 mainly for reprocessing, 7 for a roughly 50/50 split of direct disposal and reprocessing, with the remaining 11 deferring decision.

3.2.1   Recent European Commission research suggests that the maximum dose to members of the public expected from direct disposal of spent fuel is of the same order as that from the disposal of wastes arising from reprocessing. Also, as mentioned earlier, interim or long term retrievable storage will be required in all cases for fuel and wastes. Hence, from a regulatory point of view, there may not be a preference for either option on safety grounds.

3.2.2   Direct disposal is arguably not an efficient use of our energy resources, considering only about 3% of spent fuel is of no further potential value. Also, it is not clear whether the containment integrity of spent fuel directly disposed of in deep repository formations will be preserved over thousands of years. However, direct disposal does not generate the stockpiles of separated plutonium we get from reprocessing; separated plutonium has been stored safely for several decades but may present longer term containment and security problems.

3.3   To date, some 900 tonnes of plutonium have been created worldwide. The civil stockpile arising from reprocessing operations is about 150 tonnes, a significant proportion of which is stored here in the UK. Plutonium is a potential energy resource but can also be a liability requiring rigorous safeguards. There is worldwide concern about the plutonium stockpile. However, a number of countries recognise potential economic benefit in using plutonium as a fuel.

3.3.1   Existing plutonium stocks may be reduced by vitrification in glass and subsequent disposal in deep repositories, or by partitioning and transmutation (P&T), or by the manufacture of mixed oxide (MOX) fuel.

3.3.2   P&T is an expensive option for which the safety benefits are not clear. Research is currently being carried out by France, Japan, the USA and as part of the European Commission's research programme. As stated in the 1995 White Paper[1], the UK is not planning to carry out such research.

3.3.3   Belgium, France and the UK have commercial MOX fuel manufacturing capability. BNFL at Sellafield have been successfully operating a MOX Demonstration Facility (MDF) since April 1995 for overseas customers. A full scale MOX plant is currently under construction at Sellafield with planned operation by the end of 1997, again for overseas demand. There is however a limit to MOX reprocessing and at some stage, either plutonium by-products which cannot be recycled are likely to be added to the plutonium stockpile or direct disposal of fuel will replace reprocessing. Regarding UK plutonium holdings, the NII require licensees to develop transparent management strategies similar to those required for radioactive wastes.

3.4   As I stated before, repositories for the disposal of long lived radioactive materials are not currently available. The cost of a safe, well designed repository is likely to be high and may be a considerable burden or even unrealistic for countries with small nuclear industries. Also, suitable locations for repositories do not exist in all countries. The question then arises of the possibility of international repositories.

3.4.1   Would countries having their own repositories be prepared to offer a disposal service to other countries or would they allow foreign utilities to build repositories in their countries ? The operators of such repositories would achieve economies of scale whilst their customers would avoid the expense of constructing their own facilities. But we have recently seen the problems of trans-boundary shipment of wastes with the return under current contracts of HLW from France to Japan and to Germany.

3.4.2   So is it likely that we can anticipate public acceptance for the disposal of radioactive materials generated in other countries? Perhaps not, as we are all aware of the extensive opposition from locals against having repositories in their neighbourhood, even to accommodate radioactive materials originating within their own country! French and Swedish law actually prohibits the disposal of foreign wastes. The general UK policy[1] is that foreign wastes should not be accepted for ultimate disposal in the UK. However, the UK policy allows BNFL to offer waste substitution to overseas reprocessing customers under arrangements designed to ensure broad environmental neutrality for the UK.

3.4.3   Maintaining public confidence in the safety of the nuclear fuel cycle is vital to its continuance. It is particularly so for the back-end fuel cycle activities of storage and disposal. Within the UK, significant issues covering the disposal of radioactive materials are subject to consultation with interested parties which include plant operators, local authorities and regulatory bodies. This approach was adopted in the recent Government review of radioactive waste management policy. Issues associated with the UK national repository will also go through the public consultation process. Additionally, both the nuclear plant operators and the relevant statutory regulatory bodies continue to promote public awareness by, for example, the provision of public domain literature.

3.5   I would now like to give you an appreciation of UK government policy on spent fuel and radioactive waste management and its implementation through regulation.

4.   UK GOVERNMENT POLICY ON SPENT FUEL AND RADIOACTIVE WASTE

In July 1995, the UK Government published a White Paper (Cm2919)[1] reviewing its policy on radioactive waste management. The White Paper addressed spent fuel management, stating that UK Government policy is that *the question of whether to reprocess, and if so when, should be a matter for the commercial judgement of the owner of the spent fuel, subject to meeting the necessary regulatory requirements.* Regarding radioactive waste management, the White Paper requires site specific strategic planning in consultation with the relevant regulatory bodies and disposal organisations. Further details of Government policy as it affects operations will be presented to this conference in a paper by some of my colleagues entitled: "The Regulation of Storage of Radioactive Materials in the UK".

5.   REGULATORY FRAMEWORK IN THE UK

5.1   In the UK, the storage of radioactive materials on a nuclear licensed site is regulated by HM Nuclear Installations Inspectorate (NII) on behalf of the Health and Safety Executive (HSE). The regulation of disposal of radioactive waste under the Radioactive Substances Act 1993 (RSA 1993), is, since April this year, the responsibility of the Environment Agency, for England and Wales, and the Scottish Environment Protection Agency. The storage of radioactive waste on nuclear licensed sites is exempt from the requirements of RSA93, however on these sites, the NII applies an equivalent regulatory regime and there is close liaison between the regulatory bodies under the terms of Memoranda of Understanding.

5.2   The main legislation governing the safety of nuclear installations is the Health and Safety at Work etc. Act 1974 (HSWA) and the associated relevant licensing provisions included in the Nuclear Installations Act 1965 (as amended). Under the Nuclear Installations Act, a site may not be used by persons other than the Crown for installing or operating any nuclear installation unless a site licence has been granted by the Health and Safety Executive (HSE). The law allows HSE to attach any "conditions" to the site licence that may be necessary in the interests of safety, but places the responsibility for safety with the licensees of Nuclear Installations.

5.3   The site licence conditions provide a general framework for the management of safety and they enable the regulation of both materials in storage and the associated waste management practices. In particular, there are two requirements which have implications for the safety of storage in the fuel cycle. These require, in the first instance, arrangements to be made by licensees for the production and assessment of safety cases so as to justify safety at all phases in the life of a particular plant or process. Secondly, there is a requirement for arrangements to be made for the periodic and systematic review and reassessment of safety. This is very important, particularly where operations may extend or need to be extended into the future. The review must look back in time over operations to learn from experience, and also look forward to take into account mechanisms which may degrade safety. These include fatigue or corrosion of the plant. There is also a requirement to review the expected lifetime of plant and equipment in the context of strategic need, that is to determine how long does it need to do its job before the next phase of activities, ultimately leading to disposal.

6.   SAFETY PHILOSOPHY AND ASSESSMENT

6.1   In assessing the adequacy of a safety case, NII are selective in choosing what to examine. This ensures that our resources are deployed in the areas of highest risk. Our assessments recognise that no human activity is entirely free from harm. It may potentially affect health or involve risks to life. The basic principle underlying safety legislation is that risks from work activities to employees and the public should be reduced to levels which are as low as reasonably practicable (ALARP). In some cases, regulations specify particular procedures or precautions, or set limits which must not be exceeded regardless of cost. Meeting these limits, however, does not remove the duty on the employer or operator to take whatever measures are reasonably practicable in the light of current knowledge and technology to further reduce risk. The levels of risk (both individual and societal) which arise from work activities and the application of the ALARP principle are discussed at length, in HSE's publication[2], "The Tolerability of Risk from Nuclear Power Stations" (TOR).

6.2   The hazard potential from storage of radioactive material is associated with the leakage and escape of radioactivity. It is important that good engineering practice is applied to the storage of radioactive materials to ensure that these risks are adequately controlled. This potential for the loss of control of material is an important factor which must be considered in the periodic review of safety in order to demonstrate adequate levels of safety for the next period of operation, and to ensure that safety can be maintained in the context of the strategy and options for the future management of the material, whether this be imminent disposal or indefinite storage.

6.3   Risks inevitably change through the various phases of a plant's life, though each of these is still required to be ALARP. Some short term increases in risk can be acceptable and are sometimes necessary to achieve an overall reduction or elimination of future risk. For example, decommissioning of old waste storage facilities may on occasion necessitate increased short term risks with the aim of retrieving and disposing of wastes or converting them to a passively safe state to minimise further intervention. However, as always, the higher the risk and the more serious the potential consequences, the more onerous is the task of demonstrating that further precautions are not reasonably practicable.

7.   CONCLUSIONS

7.1   When spent fuel is stored prior to direct disposal, its radioactive constituents are contained within the fuel cladding. In the reprocessing option, these are separated into potentially reusable uranium and plutonium as well as low, intermediate and high level wastes. The safe storage and continued safe management of spent fuel and of the separated products are important. More generally, the issue of radioactive materials management crosses international barriers and we can learn from each others' expertise and experience. This conference provides one such opportunity. There is a need to ensure that there are ways to deal with uranium and plutonium stockpiles, the direct disposal of fuel and the disposal of radioactive wastes. These are matters of international interest and concern. Whatever solutions are proposed, we need to ensure that the options we select are demonstrably safe and allow sufficient public involvement in the decision making process; we may then have a realistic prospect of successful implementation.

I wish you all a successful conference.

8.   REFERENCES

1.   "Review of Radioactive Waste Management Policy", Cm2919, HMSO 1995.

2.   "The Tolerability of Risk from Nuclear Power Stations", HSE 1992.

A review of systems available for the storage of spent PWR fuel

G LEACH MA, DPhil, MBNES, MMineralogical Soc
British Nuclear Fuels Engineering Consultancy Services, Cheshire, UK

SYNOPSIS

A storage system for spent nuclear fuel has to fulfil certain fundamental requirements: adequate containment, shielding for gamma and neutron radiation and heat dissipation. Also, it must be robust against external hazards.

A variety of systems have now been developed which are suitable for the interim storage of spent PWR fuel. These include ponds with racks or containers and a number of dry systems based on vaults or casks. These various systems can be grouped into categories, based on the manner in which they use materials to meet the fundamental requirements. This categorisation provides a useful basis for discussion of the features of the various systems.

1 INTRODUCTION

1.1 The Need for Spent Fuel Storage

The ultimate fate of spent fuel from nuclear power plants is a matter which receives extensive discussion in the nuclear industry. Some power plant operators have decided to have their spent fuel reprocessed. Others favour the exploration of a direct disposal route. Many others are keeping an open mind, watching developments in both reprocessing and direct disposal technology. In this situation, there is an obvious need for the retention of spent fuel for an intermediate period, which could be prolonged.

The place of spent fuel storage in the nuclear fuel cycle is illustrated in Figure 1. Storage in reactor cooling pools is suitable in the short term; but these pools have only limited space and the accumulation of large inventories of spent fuel in the immediate vicinity of an operating reactor is not desirable in any case. Some extension of the storage time can be achieved by closer packing within the reactor pool, using devices for the prevention of criticality such as borate in the water and storage racks made of material containing boron. However, storage for extended periods essentially requires a separate facility. This may be at the reactor building (AR) or away from the reactor building (AFR). Clearly, extended storage of the spent fuel must be safe and must not prejudice future options for direct disposal or reprocessing.

1

1.2   The Purpose of Spent Fuel Storage

Following the above discussion, we may state that the main purpose of spent fuel storage is to retain the fuel in a safe condition, protecting workers and the Public from radiation and contamination, even in the aftermath of an external hazardous event. An important further objective is to prevent degradation of the fuel, so that future handling and processing options are not prejudiced.

1.3   Technical Requirements

Accepting the above purposes, the requirements from a storage system can be summarised thus:

a) prevention of criticality;
b) containment of radioactive material (also shields against alpha and beta emissions);
c) protection of the fuel assemblies from mechanical or chemical damage;
d) prevention of over-heating;
e) shielding of gamma-radiation;
f) shielding of neutron emissions.

The combination of items (c) and (d) ensures that the fuel will not undergo significant corrosion since external corrosive substances are excluded and a low temperature inhibits corrosion by interaction between chemical species within the spent fuel and its immediate environment.

Very briefly, (a) is provided by spacing and sometimes the use of materials which absorb neutrons; (b) and (c) are provided by physical restraints and a controlled environment; (d) is provided by adequate thermal conduction by the container and adequate heat transfer to a cooling medium and (e) and (f) are provided by sufficient thickness of surrounding material.

In addition, these protection features need to be made insensitive to external hazards: earthquake, fire, extreme weather, vehicle or aircraft impact and deliberate human disturbance. The degree of protection required has to be decided taking into account the likelihood of events and their severity: in this, local conditions must be considered.

2   GENERIC TYPES OF SYSTEM

There are a number of ways in which the requirements listed above can be met using current technology. These are described briefly in the following sections and are illustrated in Figure 2. It should be noted that in this paper the word "cask" is used to describe any thick-walled container, while the word "flask" is used to describe a container which is both thick-walled and sealed. In this paper the "primary" containment of the spent fuel is taken to be provided by the fuel cladding.

2.1   Ponds

A pond uses water to provide cooling of the spent fuel. The water also provides shielding of neutron and gamma radiation. The secondary containment is provided by the pond walls and above-pond ventilation system generally, but containers (multi-element bottles or MEBs) may also be used. Criticality is prevented by spacing, sometimes assisted by using materials containing boron for the storage racks or by adding borate to the water. Corrosion of LWR fuel is very slow in water at pond temperatures. Control of the water chemistry ensures the absence of aggressive chemicals. Protection against external hazards is provided by the large building size and strong walls.

The main disadvantages of a pond system are its size and complexity and the need for water treatment, which generates secondary waste. The generation of hydrogen by radiolysis is also a disadvantage, more particularly in the associated wet transport operations. The main advantages are the excellent cooling and shielding provided by water and the transparency of water which makes mechanical handling operations easier. Therefore, pond storage is almost universally employed at reactor sites (AR) as the first storage system for freshly-discharged fuel.

2.2 One-piece Flask

Flask storage has been developed from transport flasks. These have thick metal walls which provide shielding, containment and protection against external hazards. Many transport flasks hold the spent fuel in water but, if used for prolonged storage, this would be a disadvantage because hydrogen needs to be vented periodically. Dry flasks avoid this problem: they can be loaded under water, but the water is then removed, the flask interior is vacuum-dried and filled with a suitable cover gas which ensures an inert environment. Criticality is prevented by spacing and the use of materials containing boron for the internal partitioning within the flask. Cooling is provided by heat conduction through the cover gas and internal structure to the flask wall and then conduction through the flask wall to the outside air. Air cooling is promoted by the provision of fins on the flask exterior.

The thick metal wall provides gamma radiation shielding, but is not a good neutron absorber, so additional protection in the form of a layer of neutron absorbing material is provided. Lid seals ensure retention of the cover gas. Generally, some means of sampling the gas and replacing it are provided. The flask wall is robust and provides protection against external hazards. The high heat capacity of the flask provides resistance to fire.

2.3 Canister-in-Cask

These systems employ a metal canister which contains the fuel assemblies. The canister is stored within a concrete cask, but there is space between the canister and cask. The cask has air inlet ports at the bottom and outlet ports at the top so that air flows within it to cool the canister.

Criticality is prevented by a framework within the canister which spaces the assemblies and which may be made of material containing boron. This framework also assists in cooling by conducting heat to the wall of the canister. The canister is filled with a suitable gas and is sealed, providing containment and a non-corrosive environment. The wall of the canister is normally made of stainless steel and is fairly thin. The concrete outer cask provides shielding for both gamma and neutron radiation and protection against external hazards. The concrete cask with the canister inside can be moved within the storage site if necessary.

The canister is loaded in the reactor pool and is then drained of water, vacuum dried and finally filled with inert gas. It may have either a sealed lid system or a welded closure. There may be provision for monitoring the gas within the canister but more commonly it is completely sealed. The canister may be designed to allow it to be moved off site in a transport cask.

2.4 Canisters in a Vault

A canister-in-vault system is in principle similar to the canister-in-cask system just described, but the concrete shielding is in the form of joined walls. This can be done in a modular manner so that the vault can be extended and not constructed all at once. The principal advantage over the cask system is that shield walls are only required at the outside of the vault.

2.5 Bare Fuel Vault

In this type of system, the fuel is loaded without a container into the vault: the secondary containment is built into the vault and takes the form of storage tubes. This type of vault avoids the need for the storage canister; however, fuel needs to be transferred into the storage tubes using a sealed, shielded charge machine. The storage tubes are filled with inert gas to provide the protective chemical environment. They have sealed lids. Provision can be made for continual monitoring of the cover gas.

3   EXAMPLES OF THE STORAGE SYSTEMS

Some of the storage systems currently in use or at an advanced stage of development are introduced below and placed in context as specific examples of the generic systems described in Section 2 above.

3.1   Ponds

Pond Storage is so well established that it is not useful to describe in detail any special examples. A survey has recently been carried out by the IAEA (1) and this gives good coverage of the international scene. Much work has been carried out on the increasing of fuel storage densities and many AR pools now feature high-density racks using borated materials.

Pond storage is used away from reactor sites (AFR), sometimes as a prelude to reprocessing, as in Russia, France and Britain, and also as a long-term strategy to be followed perhaps by direct disposal, as in Sweden.

Most ponds are of the open rack type. The BNFL THORP Receipt and Storage Pond is an example of the type which uses MEBs. The Swedish AFR pond, CLAB, is an interesting example of a facility constructed underground, giving a very high degree of protection against external hazards.

One of the organisations with over 40 years experience of AFR pond storage operation is BNFL. Considerable improvements have been made as successive ponds have been brought into use (2).

3.2   Single-wall Flasks

Single-wall storage flasks have been developed from the well-tried transport flasks and therefore should all be capable of providing adequate shielding and protection from external hazards. They include the CASTOR series which is described below and other metal flasks such as the Transnuclear TN-24P and the Nuclear Assurance Corporation STC. There is also a Canadian flask which has a concrete wall.

3.2.1   *Castor series*

The CASTOR flasks are supplied by the German company GNS. Their use in Germany is described in Annexe C of reference (1). They have also found use elsewhere: an example is the use of the CASTOR V/21 flask by Virginia Power in the USA.

The CASTOR V/21 metal wall is 0.4 m thick and gives sufficient shielding for gamma radiation. Supplementary neutron shielding is achieved by incorporating polyethylene rods into the wall.

The CASTOR flasks have complex lids and seals: the V/21 model has two stainless steel lids each with both a metal and an elastomer seal. There is provision for monitoring the pressure in the space between the lids. The flask is loaded underwater and the lids are fitted. Water is then pumped out and a vacuum applied to dry out the interior of the flask. The flask is filled with helium. GNS supply equipment for the drying, including checking that residual humidity is low enough, and also for the testing for helium leakage and sampling of the cover gas during storage (3).

An interesting example for use in Europe is the CASTOR 440/84 flask (4) used for the storage of spent VVER-440 fuel at Dukovany in the Czech Republic (5). This flask can accomodate 84 assemblies - approximately 10 tonnes of spent fuel. It is constructed of ductile cast iron and is plated internally with nickel to provide a corrosion-resistant inner surface. The fuel basket is made of borated steel for criticality prevention and also incorporates aluminium spacers to enhance heat dissipation. The flask, fitted with shock absorbers, performs satisfactorily in transport impact tests.

3.2.2 *Canadian Concrete Flask*

The Canadian utility Ontario Hydro has developed a concrete-walled Dry Storage Container for CANDU fuel which fits into the category of single-wall flasks as defined here. With the co-operation of Atomic Energy of Canada Ltd (AECL), this concept has been developed for RBMK fuel (6) and it has been suggested that similar flasks could be used for VVER fuel (7). However, the concrete walls of the flask do not conduct heat as well as metal walls: this means that in order to store fuel of high heat output a great deal of steel reinforcement has to be included in the concrete to aid heat conduction and the economic advantage of concrete is then lost (8). Therefore it does not seem likely that this type of flask will become commonly used for PWR fuel; it could be used for fuel of long cooling or low irradiation.

3.3 Two-piece Flasks

The system marketed by British Nuclear Fuels and Sierra Nuclear Corporation is described below as the main example of a two-piece cask system. Mention is also made of a system designed by the French company Robatel.

3.3.1 *British Nuclear Fuels (BNFL) and Sierra Nuclear Corporation (SNC)*

The original SNC system for spent fuel storage at the reactor site was called the Ventilated Storage Cask (9) and is widely used in the USA. It has now been developed into an integrated transport and storage system known as TranStor (10). Originally designed to take 24 PWR assemblies per cask, it is now also available for BWR, VVER-440 and VVER-1000 fuel types.

The fuel assemblies are held in a canister which is made of pressure vessel grade steel and incorporates shielding in its lid. A transfer cask is provided which has bottom-opening doors and a bolted lid. The canister is loaded in the reactor pool while held in the transfer cask. Once the canister is full, the cask is moved to a decontamination area and the canister lid is fitted and welded. The canister is drained, vacuum-dried and filled with helium. The transfer cask lid is fitted and the cask is moved to a loading position. For AR storage, the canister is lowered into the concrete storage cask which is then fitted with a lid and moved to a storage position. This can be out-of-doors, and minimal surveillance is required from then on. For AFR storage, a transport cask is used to take the canister to the AFR site which is also equipped with a transfer cask and loading station.

An interesting feature is that the lid of the canister is welded shut, so the fuel is sealed in a helium atmosphere; there is no requirement for monitoring of the atmosphere or of helium leakage. The system has obtained a Generic NRC Certificate of Compliance in the USA, demonstrating regulatory confidence in the drying method and the leak tightness of the canister.

3.3.2 *Robatel*

The French company Robatel have developed a canister-in-cask system with the cask made of concrete (11). However, their literature claims a containment function for the concrete cask: it is closed with a lid and the atmosphere inside the cask (but outside the canister) can be monitored. This has the disadvantage that the heat dissipation ability is reduced, so the system is limited to fuel of over 10 years cooling and irradiation under 37 GWd/te.

3.4 Canisters in a Vault

There are a number of these available. The NUHOMS system, described first, stores the fuel in a horizontal attitude in a multi-assembly canister. The Siemens Fuelstor is also a horizontal system, but with one assembly per canister. Three vertical storage systems are also described: the Canadian MACSTOR, the French CASCAD and the British GEC-Alsthom store as supplied for the Fort St.Vrain site in the USA.

3.4.1 *Nutech Horizontal Modular Storage System (NUHOMS)*

The NUHOMS system (12) (now supplied by Vectra) is used at a number of sites in the USA. Twenty-four PWR assemblies are stored in a stainless steel canister and, for European use, designs are being sized for about 60 VVER-440 assemblies (12). The canister is welded shut, vacuum-dried and filled with helium. It is rotated into a horizontal attitude prior to being transported to the storage area on a special vehicle.

The storage building consists of a series of modules with concrete walls which provide shielding and protection. The transport vehicle incorporates a ram which pushes the canister into a storage module. Air ports at the bottom and top of each module allow air to flow within the vault to cool the canisters. There is some shielding in the canister lid to provide added protection during seal welding and canister handling operations; this also helps with shielding during the storage period, since the port by which the canister is introduced into the vault is covered by a shield door which provides less shielding than the walls.

3.4.2 *Siemens Fuelstor*

The Siemens Fuelstor also stores fuel assemblies in horizontal canisters in a vault (13). However, there is only one PWR assembly per canister. The canisters are not loaded in the reactor fuel pool: the fuel is transported to the storage facility in a dry cask and then loaded into the canisters within a hot cell facility.

The storage canisters are welded and filled with helium. They are fitted with spacers which allow them to be stacked, leaving room for air flow: the spacing also prevents criticality. The density of stacking compensates in spatial terms for the single assembly per canister.

3.4.3 *Atomic Energy of Canada Ltd. (AECL): MACSTOR*

AECL has developed a canister-in-vault system for CANDU fuel known as CANSTOR. In this, sealed canisters, each containing 60 of the comparatively small fuel bundles, are stacked on top of each other in vertical storage tubes: the tubes are not a containment barrier but hold the fuel canister stacks in position.

AECL have joined with Transnuclear Inc. of New York to extend the concept to LWR fuel under the name of MACSTOR (14). Several assemblies will be loaded into a stainless steel canister which will be fitted with a bolted sealed lid, similar to that used on metal flasks. This provides for monitoring of the helium atmosphere within the canister. The canister is lowered into a cavity in a concrete vault: no storage tube is required as the canister stands freely in a vertical position, the top of the canister being located in its port in the charge floor.

3.4.4 *SGN, France: CASCAD*

The CASCAD is also a vertical canister-in-vault design (15, 16). However, the canister, welded shut and filled with helium, is stored within a storage tube which is a containment barrier and has provisions for monitoring of the atmosphere. Thus, this system provides triple containment: cladding, canister and storage tube. The current CASCAD facility receives canisters which have been loaded at the reactor, but it would be possible to load the canisters at the storage site. It would also be possible to omit the canisters, converting the facility into a bare fuel vault.

An interesting difference between CASCAD and both the MACSTOR and MVDS systems is that in CASCAD the operation of loading the canister into the vault is performed without a shielded cask. Instead, the loading area above the storage vault is within the shielded building and the loading is carried out remotely, with no personnel in the area.

3.4.5 *Fort St.Vrain*

The Modular Vault Dry Store (MVDS) at Fort St. Vrain, USA (17) is a canister-in-vault type because the storage tubes are loaded at the reactor and transported to the store. The storage tubes have bolted lids with seals and have provision for monitoring the leak-tightness. The storage tube atmosphere is air. Although it is not a routine operation, it is possible to transfer fuel between storage tubes in the vault: when used in this mode, the system becomes similar to a bare fuel vault.

3.5 Bare Fuel Vault

The MVDS designed by GEC Alsthom Ltd. is used as the example of a bare fuel vault - that is, a vault into which the fuel is moved without a container, the containment being built into the vault.

3.5.1 *GEC-Alsthom Modular Vault Dry Store (MVDS)*

It was developed by GEC-Alsthom from the dry store at the Wylfa nuclear power plant in Wales which has been in operation for 25 years. The modern design was selected by Scottish Nuclear Ltd. when they intended to build a store at Torness (18) and a similar store is currently being constructed at the nuclear power plant at Paks in Hungary (17).

The MVDS, as applied to PWR fuel, would receive the fuel in a flask (either dry or wet) from the reactor. The fuel assemblies are removed and dried if necessary. Using a Fuel Handling Machine, each fuel assembly is loaded into a fuel storage tube which is made of carbon steel and is supported within the vault. The storage tube is sealed with a bolted lid and filled with inert gas, normally argon or nitrogen. There is provision for monitoring the storage tube atmosphere. The outsides of the storage tubes are cooled by air drawn through the vault by convection, driven by the fuel heat output. The storage tubes are spaced so as to avoid criticality, even in the abnormal event of a flooded vault.

4   DISCUSSION OF TECHNICAL ISSUES

All the systems can be engineered to meet the major requirements outlined in Section 2. However, some can meet them more simply than others and the best systems provide generally a greater margin of safety or, in other words, more robust performance. In addition, other features of technical performance differ between the systems.

4.1   Prevention of Criticality

All systems use control of spacing and some use neutron absorbers. The large vaults have an extended array with comparatively little neutron absorption between the units; they rely on the spatial arrangement. The cask systems have to limit neutron dose rates from their outer surfaces and this ensures that any array of casks will be criticality safe. Internally, canisters or flasks have criticality control by basket geometry and materials, and these features are not difficult to provide.

4.2   Containment

The manner in which containment is provided is the most obvious feature distinguishing between the systems. The fuel cladding can be regarded as providing primary containment. The secondary containment is provided in various ways: by the pond structure for a pond with racks; by the multi-element bottle in a pond/MEB system; by flask walls for metal or concrete one-piece flasks; by the canister in a canister-in-cask or canister-in-vault system, or by the storage tube in a bare fuel vault.

Some systems provide three containment barriers. In an MEB/pond system, the pond walls provide tertiary containment. The CASCAD facility also has tertiary containment if designed with containers and sealed storage tubes. The MVDS can be provided with double-walled and double-lidded storage tubes for damaged fuel assemblies.With regard to lid closures, opinions differ over the comparative merits of a welded closure and the seal systems.

4.3 Chemical Protection

Control of water chemistry provides protection of the fuel from corrosive chemicals in ponds. The low temperatures achieved ensure that water and dissolved air do not cause significant corrosion.

Inert atmospheres are used in dry stores. Zircaloy-clad fuel can be stored satisfactorily in an inert, dry atmosphere at temperatures up to nearly 400°C. There would seem to be a greater risk of leaving some moisture in with the fuel if the storage container is loaded under water. The bare fuel vault provides a period in which the fuel can lose residual moisture.

4.4 Heat Dissipation

Ponds provide the most effective cooling. Dry stores with individual fuel containers such as the bare fuel vault or the Siemens Fuelstor can provide lower temperatures than those with multi-element canisters. Heat loss from a metal flask depends on conduction through a comparatively thick wall: this is off-set by the good thermal conductivity of the metal. A closed concrete flask is not generally satisfactory for PWR fuel.

An accepted target maximum temperature is 380°C, and all the systems except the closed concrete flask can achieve this with some margin to allow for exceptional events. The differences lie in the size of this margin. As a general point one expects that the more layers of containment provided, the less efficient is the cooling.

4.5 Direct Radiation (Gamma) Shielding and Operator Doses

All systems can be designed to meet shielding requirements. The large water depth and massive walls of a pond provide good shielding, but ponds require considerable surveillance and maintenance by operators.

Dry stores require less attention. Vaults easily fulfil the shielding requirements because of their thick outer walls. Bare fuel vaults will produce some direct dose penalty because of the large number of fuel handling operations and maintenance of the handling machine. Casks and flasks have the disadvantage that every wall requires external shielding standards while the weight has to be kept to a reasonable minimum if the advantage of moveability is not to be lost. The canister-in-cask avoids the conflict in flask design between the better shielding but lower heat dissipation of a thicker wall.

4.6 Neutron Shielding

A pond is the best system because of the good neutron absorption and large thickness of water. A vault has an advantage over casks or flasks. Concrete casks have an advantage over metal because concrete is a better neutron shield: metal flasks require special provision for neutron shielding and this thickens the walls, which is a disadvantage for heat loss.

4.7 Passivity

An ideal system is "passive", requiring no services or monitoring. The welded canister is a passive system. Lidded flasks and storage tubes require monitoring for gas leakage; occasional seal replacement may be necessary. A vault has an advantage over ventilated casks because the larger air ports require less surveillance. Ponds require maintenance and water monitoring.

4.8 Monitorability

It is an advantage, if containment becomes damaged, to be able to detect this at an early stage. For example, the monitoring of the atmosphere within the storage tubes of a Modular Vault Dry Store could enable the deterioration of cladding to be detected. On the other hand, systems with welded canisters do not permit detection of cladding failure and therefore, if such failure occurred, there would be only a single barrier between the radioactive contamination and the environment and this situation would not be revealed.

Ponds provide for monitoring; MEBs, if present, limit the spread of contamination. The CASCAD system, with three barriers including monitoring between the second and third, offers a good combination, with sealing of the inner containment but monitoring for safety reasons; but it does not provide a means of detecting cladding perforation. Another consideration is the number of assemblies stored together. In a multi-assembly container, one cladding failure could contaminate, many assemblies, whereas with one assembly per container, only the one would be contaminated.

4.9 Secondary Waste and Decommissioning

Pond water cleaning generates secondary waste which requires treatment and disposal; MEBs assist by controlling the spread of any contamination, but the MEB eventually becomes a waste item for disposal. The large pond structure eventually will require decommissioning.

Dry storage does not generate secondary waste during operation but the facility will require decommissioning. The eventual quantity of radioactive waste is likely to be minimised by the use of moveable multi-assembly canisters, since the building structures and handling equipment are kept clean and do not become radioactive waste.

4.10 Moveability

Metal flasks are designed for both transport and storage, and so provide the opportunity for quick removal. A canister-in-cask can be moved on site, but for a longer journey the canister must be moved in a transport cask. With ponds, MEBs make the fuel more moveable. Movement of fuel in a pond rack or bare fuel vault depends on a fuel handling machine. The individual canisters in the Siemens Fuelstor would require more handling operations to move the fuel than a multi-assembly canister system.

4.11 Fire Resistance

Ponds provide excellent protection. Vaults are very good because of their size. A single-wall flask is superior to a canister-in-cask, in which hot gas from a fire could flow inside the concrete cask and affect the canister which is of comparatively light construction, whereas a flask is massive and enclosed and provides a considerable delay in the heating of the contents.

4.12 Impact Resistance

The massive structure of a pond or vault store provides very good impact protection. However, all the systems are very robust.

4.13 Earthquake Resistance

The metal flask is inherently the best design against damage by earthquakes. Large buildings are more susceptible to damage than cask-based systems. The low profile of the NUHOMS vault is a good feature. Earthquakes occurring during the lifting of bare fuel in a vault or pond may be the greatest hazard. However all the systems can be designed with earthquake resistance.

4.14 Wind Resistance

All the systems are resistant to high winds during the storage phase. Debris blown by wind might block the air ports of the canister-in-cask or the NUHOMS vault; but these could be easily cleared. The outdoor operation of loading the canister into the NUHOMS vault may be interrupted by sudden high wind.

4.15 Flood

It is conceivable that flood water could block the air inlet ports of the canister-in-cask or the NUHOMS vault. Mud and debris carried by flood water could block the air inlet ports of ventilated concrete casks; therefore this kind of store should be located in an area free of severe flooding: in practice, this is not a significant restriction.

4.16 Cold Weather, Ice and Snow

All systems will be resistant to cold weather and snow. Metal casks and canisters within concrete casks have been selected to be resistant to low temperature, though handling may be restricted. The warmth from the stored fuel is expected to be sufficient to prevent any prolonged blockage of the air inlet ports of the canister-in-cask by snow or ice.

4.17 Hot Weather and Sunshine

Very hot weather and strong sunlight could reduce the cooling ability of dry storage systems. The canister-in-cask should perform better than the metal flask because there is separation between the sun-heated outer part and the fuel container. However, all systems can cope with these conditions.

4.18 Animals & Birds

The buildings of the pond and vault systems make these very resistant to any ill effects caused by animals and birds. The vents on the dry storage vaults are fitted with screens and are of a large size which will make them resistant to blockage by animal carcasses, etc. The canister-in-cask air inlets could be blocked by insects but can be inspected and cleaned. The strong draught and elevated temperature in the space between the canister and concrete cask wall would prevent the space being occluded by spiders' webs.

4.19 Human Disturbance; Safeguards

All the systems are secure and resistant to simple malicious damage. The fuel handling machines in ponds or bare fuel vaults are probably the most sensitive items, but access to them is difficult. Flasks or casks standing outside are the most sensitive to embarrassment by protest action. In the event of military action, no system is proof against a substantial attack, but the buildings are more resistant than the flasks and casks to smaller weapons. A prolonged loss of services or inability to take remedial action could cause loss of water chemistry control in a pond.

The diversion of spent fuel to produce weapons is an important issue, but all the storage systems can be provided with recording methods to prevent this. There is a degree of added protection if the spent fuel is difficult to move (c.f. section 4.10). In practice, spent PWR fuel would be an extremely difficult starting material for the production of a weapon.

5   CONCLUSION

There are a number of systems which can adequately provide storage for spent PWR fuel. These can be grouped into categories which reflect the different ways in which they achieve the fundamental requirements. These different categories show differences in other characteristics, which may be more or less important to different potential customers. No particular category or system could be said to be superior in all respects. Selection of a specific system thus depends on the particular requirements of the customer and on the cost of the system. A potential customer has plenty of choice and needs to assess the choice on the basis of detailed quantitative technical performance, cost and the compatibility of the sytems with his own spent fuel management strategy.

REFERENCES

Conference Proceedings

A. P.-E.Ahlstroem et al. (eds). "High Level Radioactive Waste and Spent Fuel Management". Volume 2 in: J.Marek, R.Kohout et al. (eds) "Nuclear Waste Management and Environmental Remediation". Proceedings of conference organised by the American Society of Mechanical Engineers, Czech and Slovak Mechanical Engineering Societies and the Czech Nuclear Society; Prague, 1993.

B. "The Prospects for Dry Fuel Storage". Proceedings of conference organised by the Institution of Nuclear Engineers; London, 10th November 1993.

List of References

1. Guidebook on Spent Fuel Storage. IAEA Tech. Report 240; 2nd edn., 1991.

2. BNFL Experience in the Design and Operation of Spent Fuel Storage Facilities. C.Bristol; Project Management Department, BNFL Engineering Division; pp 253 - 257 in [2358].

3. Handling Systems for Transport and Storage Casks. Brochure by GNS.

4. CASTOR Cask with High Loading Capacity for Transport and Storage of VVER-440 Spent Fuel. R.Diersch, D.Methling & G.Milde; GNB, Germany; pp 45 - 50 in [2358].

5. Interim Storage of Spent Fuel Assemblies from VVER Reactors, using as an example Cask Dry Storage for the Czech Dukovany NPP. W.Botzem, P.Arntzen, R.Diersch, R.Laug, V.Reznik & P.Cech; NUKEM, GNB & CEZ; pp 595 - 599 in [2358].

6. Development and Prospects of Canadian Technology for Dry Storage of Used Nuclear Fuel. P.D.Stevens-Guille & F.E.Pare; Ontario Hydro & Atomic Energy of Canada Ltd. (AECL); [2466] session 1, paper 4.

7. Dry Store for VVER/RBMK Fuel Using Reinforced Concrete Containers. W.F.Howe, L.Grande & P.J.Armstrong; Ontario Hydro, Canada; pp 545 - 548 in [2358].

8. Transport, Storage and Disposal in the Federal Republic of Germany. K.Janberg Gesellschaft fur Nuklear-Service mbH (GNS) & Ges. fur Nuklear-Behalter mbH (GNB); [2466] session 2, paper 1.

9. Ventilated Storage Cask System. Paper published by Sierra Nuclear Corporation; January 1993.

10. Bridging the Gap - the Interim Storage Solution. P.J.Roberts & R.A.Warren; Paper in this conference.

11. Brochures by the Robatel company, France.

12. Background on the NUHOMS System and its Current Status. J.C.Ritchie; Pacific Nuclear Systems Inc., USA; [2466] session 2, paper 3.

13. Use of "Fuelstor" Vault for Interim Spent Fuel Storage in Czech Republic. H.Gunther & Z.Valvoda; Siemens, Germany & DIAMO, Czech Republic. pp 573 - 585 in [2358].

14. "MACSTOR": Spent Fuel Storage for the Nuclear Power Industry. F.E.Pare, P.Pattantyus & A.S.Hanson; AECL, Canada & Transnuclear, New York, USA. pp 275 - 279 in [2358].

15. The "CASCAD" Spent Fuel Dry Storage Facility. G.Bonnet & M.Giorgio; SGN, France; pp 565 - 571 in [2358].

16. SGN Dry Vault Storage in France and its Prospects. C.Bonnet & G.Ducroux; Societe Generale pour les Techniques Nouvelles (SGN); [2466] session 1, paper 3.

17. Experience and Applications of the GEC Alsthom Modular Vault Dry Store. C.Ealing; GEC Alsthom Engineering Systems; [2466] session 1, paper 1.

18. Progress with Scottish Nuclear Limited's Dry Fuel Store Proposals. I.S.Cathro; Scottish Nuclear; [2466] session 1, paper 2.

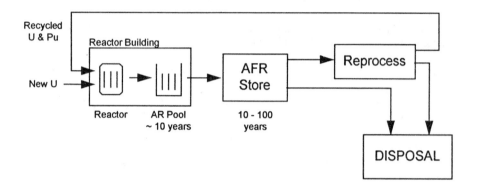

Fig 1 Place of AFR Storage in spent Fuel Management

Types of Pond

Open racks

Multi-Element Bottles

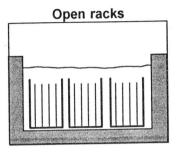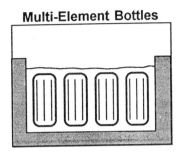

Types of Cask

Single-Wall Flask

Canister -in- Cask

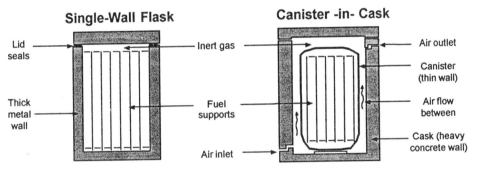

Lid seals — Inert gas

Air outlet

Canister (thin wall)

Thick metal wall — Fuel supports

Air flow between

Air inlet

Cask (heavy concrete wall)

Types of Vault

Canister -in- Vault

Bare Fuel Vault

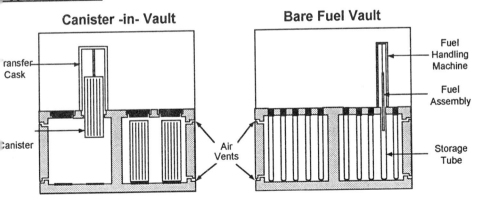

Transfer Cask

Canister

Air Vents

Fuel Handling Machine

Fuel Assembly

Storage Tube

Fig2 Storage Options

A systematic approach to decision making for spent fuel management

BAILLIF
USYS, France
A BROWN BSc, MSc
Consultant, London, UK
TAKÁTS BSc, MSc, ENS
International Atomic Energy Agency, Vienna, Austria

Synopsis

This paper gives a brief resume of the three important options for spent fuel management. These are: reprocessing, direct disposal and interim storage. It is suggested that interim storage is a legitimate option that involves a conscious decision of deferral with respect to reprocessing or disposal. A systematic technique is presented which involves careful analysis of all valid factors that influence a decision. The factors or attributes are given weighing in orders of importance, and techniques such as multi-attribute analysis can be used to derive a preferred decision. It is recognized that there are other outside influences that can affect an optimum decision but it is recommended that all countries adopt a systematic decision making process or review their current decisions.

1. INTRODUCTION

The operation of nuclear power plants (NPP) generates spent nuclear fuel for which suitable management arrangements must be made. Worldwide there are more than 400 nuclear power reactors in operation, producing around 17% of the world's electricity.

By the year 2000, about 225 000 tones of spent nuclear fuel will have been discharged from these power plants. While most of this material will be stored for an extended period prior to final disposal, a significant amount is already committed to being reprocessed, thereby reducing the accumulated quantity of spent fuel that will require interim storage and eventual disposal.

2. THE AVAILABLE ALTERNATIVES FOR SPENT FUEL MANAGEMENT

There are basically three options available for a country to follow in the area of spent fuel management. The options are illustrated in Fig. 1. These are discussed further as follows:

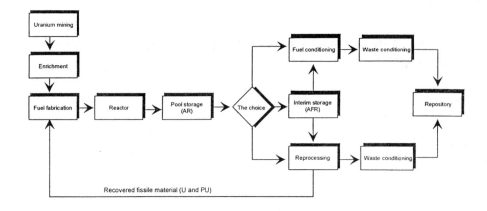

Fig. 1 : *Options for spent fuel management
(AR - At Reactor, AFR - Away From Reactor)*

2.1 The closed fuel cycle

For this option the spent fuel is transported to a reprocessing facility where the residual fuel values (uranium and plutonium) are separated from fission products and other actinides. The recovered uranium and plutonium are returned to the fuel cycle for re-use in subsequent fuel fabrication as uranium or uranium/plutonium mixed oxide (MOX) fuel. The fission products and remaining actinides as well as any medium and low level wastes from reprocessing must be solidified and encapsulated or otherwise processed for disposal.

2.2 The once through fuel cycle

In this case, the spent fuel is transported to the final disposal facility where it is suitably conditioned and packaged for the disposal site environment. It is placed in the repository without any recovery of residual fuel values.

Direct disposal of the spent nuclear fuel into a geological repository without recovery and recycling of the residual fuel values obviates the need for capital investment in reprocessing and high-level waste facilities although there will be a similar high cost for developing a repository and licensing it. The direct disposal choice requires a programme to develop and operate the geological repository. Many of the necessary fuel conditioning processes are available.

2.3 Deferral

Selection of deferral means that the utility presently does not wish to decide which of the previous two courses of action, viz. reprocessing or direct disposal, to adopt at the present time. This delay should be a conscious policy decision, since it implies interim storage. Sometimes this occurs by default but it is preferable to make a deliberate choice, because this allows proper plans to be made before the available storage capacity at the plant(s) is used up. These plans could include removal of the spent fuel from the at-reactor (AR) storage and its transfer to a long term interim away-from-reactor (AFR) storage facility where it would be retained until future policies on spent fuel management are established.

3. MAKING A CHOICE

A generalized methodology for deciding which of the spent fuel management alternatives constitutes an optimum is outlined briefly in this paper. More details can be found in the IAEA Technical Report Options, Experience and Trends in Spent Fuel Management, (Reference 1). The particular conditions in a given country will be incorporated in this approach to the best possible extent. A suggested set of characteristics or attributes for use in the decision making methodology is presented and discussed briefly. There might be also other overriding political factors in the decision making process which are individual for a given country; they are not included here. As has been discussed in Section 2, all options include an element of storage, although the selection of a particular storage system, its location and timing have to be made.

3.1. Evaluation and selection

There are no general rules for decision making on spent nuclear fuel management options but there are some basic guidelines that are useful to follow. Fig. 2 should be considered as a basic list of steps to be taken in the process of spent fuel management option selection. The steps are as follows:

3.1.1 Projected Nuclear Power Plant Programme

A realistic assessment of the country's future nuclear power programme is important because this will influence the policies and economics of the fuel management choice. For example, countries with small initial nuclear programmes might find it more economic to defer for a period until the size of their programmes justify a decision. High and low and best estimates are needed to take account of variable future growth plans. Most countries will already be considering future energy scenarios and should be able to make year by year projections of Gwh(e) of nuclear power. The type of plant and predicted availability must also be estimated. It is realized that the future demand for electrical power cannot be accurately established and consequently, any preferred option should have enough inherent flexibility to adapt to programme change.

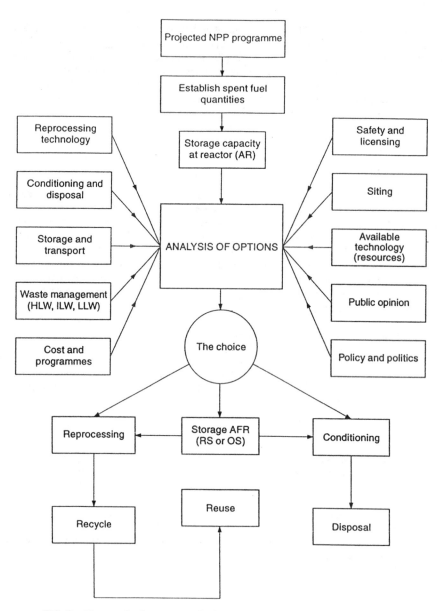

FIG. 2. *Elements in the process of selection of a spent fuel management option.*

3.1.2 Estimates of spent fuel discharged from all NPPs

This quantity can be derived from the above estimate of electricity demand but will only reflect the actual quantity of fuel discharged from reactors based on Gwh(e). There are a number of factors that will determine the actual quantity to be handled by the back end of the fuel cycle. These are: the actual burnup achieved, the operating cycles and performance of the plants and the decisions reached in any country on the amount of fuel to be reprocessed and recycled. There is also fuel which cannot be or will not be reprocessed, e.g. experimental fuel.

Generally a model is used to define the amount of spent fuel applicable to the options being evaluated and this must be reviewed during the decision making process and subsequently when an option is being implemented.

3.2 Analysis of options

In order to analyse options in a systematic way it is necessary to evaluate the characteristics or attributes of all the elements that make up an option. Attributes are the characteristics of a particular option such as costs, state of technical development, radiological risks etc. An attribute can have a number of elements which are brought together to evaluate it. When the attribute has been quantified, it has a numerical value which is called a parameter. It is suggested that these parameters can be considered in 10 elements which are shown in Figure 2 and are discussed as follows:

3.2.1 Reprocessing Technology

An understanding of reprocessing technology and the supporting R&D programs is important in considering this option even if the country does not intend to develop the facility itself. If services are sought in other countries, then at least the full implications and costs and the arrangements for waste disposal and the return of recycled material should be appreciated. The quantities of spent fuel that may be sent for reprocessing must be established here and also the needs for recycled material, e.g. MOX fuel.

Regarding fuel that is suitable for reprocessing, there may be a range of quantities to consider giving an optimum solution and at least the scenarios of 0% and 100% committed to reprocessing should be considered, to bound the range of this option. The available world reprocessing capacity must be considered if international services are to be used. Where a country wishes to develop its own reprocessing services, all the facilities that have to be implemented to allow this complex activity will have to be identified and costed. For reprocessing services contracted with other countries, facilities required to store the wastes and safeguard the recycled materials produced by the reprocessing activities will have to be identified and costed by the spent fuel owners. The reuse of recycled material for MOX or normal fuel is part of the overall costing of this option.

3.2.2 Conditioning and direct disposal of spent fuel

This is a field that is still being developed in a number of countries and the IAEA has produced a Technical Report on the subject of conditioning (Reference 2).

It is important to consider this area of developing technology when decisions are being made on interim storage. There is an important interface between storage and disposal here which, if considered carefully, can avoid future problems and costs, i.e. at least in the unnecessary rehandling of fuel and avoiding the generation of additional waste. To assist in the decision making process, it is essential that the costs and timescales for conditioning and final disposal of spent fuel are estimated even if these are only tentative at this stage. These costs will depend on the quantity of spent fuel destined for disposal and these estimates will need to be adjusted according to the amount of fuel reprocessed and/or recycled for each option as discussed above.

3.2.3 Transport and Storage

Although transport and storage are unavoidable activities, they must be considered in the decision making process because of the implications on other factors such as costs, radiological risks, available sites, public opinion etc. Capacity and cost estimates per unit volume for storage and transport activities should be estimated for each option. The quantity of fuel to be stored at any time will depend on the plant pool capacities in the short term and on the reprocessing and disposal options in the longer term.

An important parameter will be the availability of the desired at-reactor (AR) storage capacity and the capacity of various types of away-from-reactor (AFR) storage. In particular, storage systems must be flexible enough to accommodate changes in other parts of the fuel cycle without serious implications on costs, transport arrangements or safety. An important consideration in selecting a fuel management approach or a spent fuel storage concept is how well the proposed approach interfaces with existing facilities especially at current or even future power plants and also between AR and AFR storage systems. Fuel will eventually have to be retrieved from storage systems and interfaces with reprocessing plant and/or repositories must be considered.

3.2.4 Waste Management

The fuel cycle produces all types of waste and the quantities produced by each option will vary. The particular quantities will need to be estimated for each option in order to produce preliminary designs of waste treatment plant to derive costs, storage and disposal requirements so that valid comparisons can be made. All storage facilities produce low- and intermediate level waste (LLW & ILW) to a greater or lesser extent although the overall quantities should not be large. For the reprocessing option, the most important waste stream will be the high level waste (HLW). This waste will generally be conditioned as part of the reprocessing operation but may be costed separately.

3.2.5 Costs and Programmes

Cost estimates should be made for all the elements of the three options allowing for unknowns where necessary. The main cost elements will include reprocessing, waste management, storage, transport, conditioning and disposal costs. The development costs for new technologies and also the cost of public awareness campaigns has to be included. Allowances and contingency margins should be clearly identified to avoid their being compounded in comparing options. The costs of all the various elements of the different options are then brought together with the expected time scales so that a meaningful comparison can be made between the options using discounted cash flow techniques. In order to obtain a true comparison of costs between options it is important that total life-cycle costs are established. This requires a good understanding of all the complex processes and systems described above. When comparing life-cycle costs it is important to have a common time base and to calculate net present value of all costs at a reference date. The levels of initial investment will vary significantly between options and immediate or early cash requirements could be an important factor. Costs, of course, are not necessarily the most important factors because, in the end, a wide range of issues will influence decisions in a country.

3.2.6 Safety and Licensing

The important activities of the three fuel management options should be identified and the radiological risks to the public and the workers should be assessed in order to determine the preferred option in terms of safety. It is an internationally recognized recommendation that for any nuclear activity, the process is justified, doses are limited, protection and safety are optimized. Estimates of potential radiological exposures can be obtained from a knowledge of the processes and the inventory of radioactive materials and the risk criteria and regulations in a country. The radiological risks associated with transport and storage periods must also be taken into account. The development, design and licensing of a reprocessing plant may take 10 years or more while the work for site characterization and development of a first repository for HLW and spent fuel has been ongoing for longer periods of time in a number of countries. A licence application for final disposal of this type of waste has not been filed yet, whereas repositories for non-heat generating waste (ILW+LLW) are already in operation. However at this stage of the decision making process, it should be assumed that licensing will be possible for all installations that are selected from proven technology although the degree of difficulty in gaining acceptance will vary from option to option. Adequate allowance should be made for uncertainties in costs and time scales.

3.2.7 Siting

The availability of suitable sites within a country and their potential for licensing and gaining public acceptance must also be ascertained. Clearly unlicensable sites should be rejected. The ease (or difficulty) of obtaining sites for spent fuel management facilities within the country may become the determining factor in deciding which course to follow. An NPP or reprocessing plant may also remain a licensed site for a very long period before it is totally decommissioned. If the intended facility is to be located on an existing licensed site which has been accepted by the local population, then siting of the new facility may be easier. From the above, it can be expected that extensions to existing AR storage and any new AFR storage on the site will be easier than gaining acceptance for AFR storage at a completely new site. However, extending facilities on the site, which interact with the reactor itself must take into account the implications of eventual plant decommissioning. Another important siting criterion is geographical location within the country in order to minimize spent fuel and waste shipments. Account should be taken of future development plans within the country.

3.2.8 Available Technological Resources

The management of spent fuel can require advanced technology and a country must assess whether it has indigenous resource for this, whether it wishes to develop this expertise itself or whether it seeks developed technology and expertise from other countries which have licensed facilities and can offer these at competitive prices. The acquisition of disposal technology can be deferred for a period provided safe and adequate interim storage is provided. Interim storage systems are available on the international market. Eventually each country must develop its own disposal route and must provide sufficient technical resource to prepare for this in good time.

3.2.9 Public Opinion

The safe, economic and effective management of spent fuel is absolutely essential for the development of nuclear power in any country and public acceptance is vital. There are many instances where effective spent fuel management activities are being hampered by adverse public opinion when clearly safe courses of action should and can be taken. The quantification of this attribute may prove difficult and speculative. It is an attribute that can be left to one side initially while purely technical issues are evaluated and then considered later in the decision making process.

3.2.10 Policy and Politics

The country must establish a policy within its political and legislative system to allow the recommendations from any decision making process to be implemented. This process will take account of standards, licensing, siting, costs, safety, public opinion, etc. It would be advisable to set up an independent committee to advise on difficult technical and nuclear matters. Quite often in countries, political issues are overriding and although this is unavoidable and will influence decisions strongly, it is nevertheless important to have conducted a systematic decision making process. An optimum solution based on quantifiable facts is then available to policy makers and this can be a basis for political decisions. The application of policy within a country is a decision making attribute which should be considered separately from the more quantifiable attributes. A policy of deferral may look attractive but there must be an awareness of future commitments and provisions must be identified to cover the total lifetime costs of spent fuel management and to provide the necessary future funding.

3.3 Analysis of Attributes

An analysis of the attributes that will be generated from the above will be complex and not all the tems will be easily quantifiable. It may be useful to separate those which are quantifiable, eg. costs, time scales, radiological risks, waste and spent fuel quantities etc. from those which are not, or for other valid reasons should be set to one side from the initial decision analysis, eg. government policy, public opinion, available sites etc. It should be noted that the quantity of spent fuel passing through the cycle will vary according to the amount of fuel committed to reprocessing and adjustments will have to be made to correct for this. A spent fuel flow model can be set up to perform these analyses.

The selection of the most appropriate technique or method is generally based on the level of detailed information that has been collected for the different options. Obviously better quality information will yield a more robust decision. A well-established method for analysing the data is called Multi-Attribute analysis or sometimes Multi-Criteria analysis. The principle is to analyse the different options in terms of their attributes; these being the particular characteristics which distinguish them from other options. In order to compare the different attributes in a meaningful way, they need to be quantified as numerical parameters. First of all, the attributes need to be ranked according to importance or merit. If the attribute has a numerical value, e.g. a cost or a radiological risk etc., then the parameter is readily usable but if attributes are not easily quantifiable, then they can be ranked on a scale of say 0%- 100% according to weighing and relative importance. In this way numerical parameters can be obtained for all the attributes and useful and valid comparisons can be made.

Additional weighing is needed between groups of attributes, e.g. cost and program, safety and radiological risk, state of technical development, licensing potential, site requirements etc. The summation of the weighted values of the parameters determines the score for each option. Note that there could be sub-options within reprocessing in terms of different quantities of fuel committed to reprocessing, e.g. 0%, 20%, 50% etc. In making use of the Multi-Attribute analysis technique, special attention is required to establish the hierarchical structure that defines the weights to be assigned to the various parameters or groups of parameters. These should not be arbitrary but should be chosen systematically. A consensus agreement on the weighing values is vital for a decision that will have general acceptance within a country and these weighing values should be chosen by a group of experts at a formal meeting or debate with representatives covering a broad spectrum of the country's interests. Sensitivity analysis should be applied to the results to determine whether the outcome is robust and will withstand adverse criticism. For example if the result can withstand a large weighing on safety and a low weighing on cost then it is more likely to gain public acceptance.

When the scores for each option have been generated and the initial preferred option is identified, it will be necessary to apply the other less tangible factors and those which are more subjective or did not have the distinguishing features needed for the initial analysis. These are factors such as Government Policy, Public opinion, siting problems and a number of local and national constraints. These may alter or modify the chosen option but, at least, the initial decision is a sound reference point based on a systematic approach which considered safety, costs, technical factors etc.

3.4 The Choice

From the three possible alternatives, one choice will emerge from the decision making process. The alternatives are:

- A decision to reprocess a part or all of the fuel with appropriate storage,
- A decision to plan only for direct disposal with interim storage,
- A decision to keep options open using interim storage (deferral).

All these are perfectly valid but it must be remembered that disposal of essentially all the material will eventually be necessary. It is known that many countries have indicated their decision. Some are definitely only pursuing direct disposal, e.g. USA, Sweden, Canada, while others are keeping options open and yet others are taking advantage of available reprocessing capacity eg. France, Germany, Japan, Switzerland and UK. In many countries the decision was arrived at by historical, legislative or strategic reasons. Today the options have to be evaluated in view of changes in the world situation particularly in the slow down of Fast Breeder Reactor programmes, the low cost supplies of raw uranium and also the intrinsic value of reprocessing in reducing volumes of spent fuel and the use of MOX together with the much more efficient and effective use of fissile uranium and plutonium.

4. CONCLUSIONS

This Paper has explored the field of spent fuel management from reactor discharge to conditioning for final disposal. It is a complex field of nuclear technology and is receiving an increasing amount of attention as the stocks of spent fuel increase worldwide. Reprocessing and direct disposal of spent fuel are contentious areas both within the nuclear power industry and in the public domain. Storage technology however has matured rapidly over the last ten years and this will give the Industry the opportunity to reassess the options available for spent fuel management. At present reprocessing capacity constitutes between 30% and 40% of the total annual spent fuel arisings. It is an important part of spent fuel management. It is desirable that the technology is preserved to continue to provide opportunity for recycle in the coming decades. In the meanwhile, reprocessing and direct disposal must be addressed and reviewed periodically as the world situation continues to change.

Finally it is recommended that countries with well developed or emerging nuclear programmes should embark on a formal decision making process that has a sound technical basis. Deferral should be viewed as a conscious decision and not one of default.

References

(1) INTERNATIONAL ATOMIC ENERGY AGENCY, Options, Experience and Trends in Spent Fuel Management, IAEA Technical Report Series No. 378, IAEA, Vienna (1995)

(2) INTERNATIONAL ATOMIC ENERGY AGENCY, Concepts for the Conditioning of Spent Nuclear Fuel for Final Waste Disposal, IAEA Technical Report Series No. 345, IAEA, Vienna (1992)

e first US dual-purpose cask using NAC transportable orage technology

STUART US-ANS
C International Inc., USA
EREZ de HEREDIA
ipos Nucleares SA (ENSA), Spain

ABSTRACT

In 1994, a significant event occurred in the technology of spent fuel management systems when the U.S. Nuclear Regulatory Commission (USNRC) issued a transportation license for the first U.S. dual-purpose cask, the NAC-STC, pioneered and designed by NAC International, Atlanta, Georgia. Several unique design features allow the cask to achieve a high capacity equal to 11 MTU of spent fuel. These features include multiple stainless steel shells; high-purity lead for gamma radiation shielding; a very efficient neutron shielding material known by its trade name of NS4FR; and an internal basket that is both robust and heat transfer efficient. The cask also incorporates a double-lid concept wherein the inner lid is identified as part of the cask containment or confinement shell. The outer lid provides an impact protector, shielding for operators, and a testable, monitorable leakage control volume. This outer lid may ultimately be qualified as an alternate containment boundary, if needed to extend the cask life, and be utilized to accomplish a final permanent welded seal for the fuel prior to burial in its unopened container.

1.0 The NRC Approved Package (NAC-STC)

On September 30, 1994, the U.S. Nuclear Regulatory Commission (NRC) issued the transportation Certificate of Compliance (C of C) for the NAC-STC, the first Transportable Storage Cask. A few months later, on July 19, 1995 the storage approval was also issued by the NRC. These approvals were supplemented by the U.S. Department of Transportation (DOT) issuance of an IAEA Certification for the same package. The package is shown in figure 1. The key features of the package are:

- A steel/lead/steel/NS4FR wall about 30cm thick

- Cooling fins of copper/steel embedded in the NS4FR[1]

- Overall dimensions of 483cm long/249cm diameter

- Internal cavity 419cm long/180cm diameter

- Wood impact limiters with steel skins

The key listing of allowable contents taken from the C of C is shown in figure 2. Essentially, all of the PWR fuel types are permitted, including the Optimized Fuel Assembly (OFA). As noted in the C of C, the key limits are:

- Maximum of 26 PWR fuel assemblies of the type listed

---

[1] NS4FR is a trade name for a highly efficient neutron absorbing shielding material made from a mixture of hydrocarbon resin materials. The technology for this material is owned by Genden Engineering Services and Construction Company (GESC) of Japan, who has licensed it to NAC for sale outside Japan.

- 4.2% maximum initial enrichment of U^{235} weight percent

- Maximum burnup of 40,000 MWD/MTU with at least 6.5 years cooling

- Maximum burnup of 45,000 MWD/MTU with 10 years cooling

- No burnup reactivity reduction credit is assumed

The cask has a removable basket and has two lids, the inner lid (23cm thick) being designated as containment. The inner lid has two metallic o-ring seals and the outer lid (13cm thick) has a single metallic o-ring seal. The cask cavity can be sampled while maintaining containment; the interseal volume between o-rings of the inner lid can be monitored; and the interlid volume between the lids can also be monitored. The cask is filled with helium initially at one atmosphere absolute pressure with the interlid region pressurized to 7.7 atmospheres absolute and thus acting as an overpressure seal for leakage control.

The steel/lead/steel outer body is not new to the NRC, as two such NAC casks are storing fuel at the Virginia Power Surry site in Virginia. After conducting appropriate drop testing in accordance with 10CFR71 and IAEA standards, the NRC fairly readily accepted the cask body and turned its attention to the interior basket. No basket with the structural payload and heat dissipating capability demonstrated in the application had previously been approved by the NRC. Figure 3 shows a cut-away sketch of the basket (actually a 21 PWR German fuel version). This tube and disk concept has been patented by NAC. The disks are of two types. One is made of a type of stainless steel known as 17-4 PH, used for its structural capacity. There are 31 of these. 13mm thick each. The other disks are referred to as aluminum heat fins, of which there are 20 at 16mm thickness using 6061-T6 aluminum, an ASTM material which has been used in the U.S. space program. The tubes, one for each fuel assembly, are made of 304 stainless steel having an inner and outer shell with a sheet of Boral encased between them. The Boral is on all four sides of the tube and is the full length of the active fuel. No structural requirements are placed on the tubes. And no reactivity reduction credit is taken for the fuel (i.e., no burnup credit).

The drop-testing program used a 1/4-scale model as shown in the figure 4 that included a scale-model basket and dummy fuel assemblies. The model was fabricated at ENSA in Spain and the tests were conducted at the Winfrith facility of AEA in southern England in 1990 and 1991. A series of 1/8-scale impact limiter tests were also conducted to develop an analytical method of predicting crush characteristics of wood. This same 1/4-scale test series has since been duplicated (in 1993) for a slightly different size STC cask designed for the Trillo plant in Spain where somewhat longer and larger German fuel is used. These additional tests were performed successfully and further demonstrate the robust features of the NAC steel/lead/steel technology as well as the tube and disk basket design.

The thermal properties of the basket received extensive attention during the NRC review process. The focus here was the need to demonstrate clearly that the heat transfer was efficient and correspondingly the peak fuel cladding temperature was within limits. Closely related, however, are the structural properties of the basket which are directly related to the temperature of the materials. Thus the thermal and structural considerations of the basket design were closely related.

Figures 5 and 6 show the detailed thermal mapping that was performed by NAC engineers to demonstrate not only the performance of the basket, but also the ability to postulate design deficiencies or operations-induced malfunctions and examine any effects on the thermal efficiency of the basket. Various NRC "what if" scenarios, such as no heat transfer within certain components, proved that the decay heat of the fuel is transported effectively in three dimensions in the basket neutralizing the peaking factor effect generated within the fuel during its power history. As a result, the inner surface of the cask is exposed to a fairly uniform flux of heat across its entire surface including the lids and bottom forging. Transference of the heat through the metal cask wall to the outside atmosphere is thereafter fairly straightforward. The net result is that even if all of the fuel loaded in the cask were to be at the maximum design conditions, peak fuel temperature is 316°C which is well below the NRC limit of 360°C.

In most practical circumstances, the fuel loaded will not all be design bases fuel and so actual peak temperatures will be even lower with additional margin to the limit.

2.0 Significant Events In The Licensing Process

NAC has over 25 years of experience licensing transport and storage packages to NRC requirements. The review, analysis and documentation required for the licensing of the NAC-STC is by far the most comprehensive and rigorous we have seen. Some of the significant events of this process:

1.   There are two lids on the cask. The original basis was the inner lid would serve as confinement in the storage mode and the outer lid as containment in the transport mode. Consequently, all drop tests were conducted with only the outer lid performing containment. The inner lid was installed but not leak tight, i.e., it was bypassed. This allowed a demonstration of the acceptability of the outer lid for transport containment. Although this configuration passed all of the drop tests, NRC rejected the concept that the containment boundary definition would change according to the mode of use of the cask. Only one lid was allowed to be designated containment in all cask modes. NAC selected the inner lid as is documented in the license. The outer lid is still a part of the design and it has been proven technically to be accident resistant. In addition, it provides a second monitorable volume to assure leaktightness of the package. In the future, this second lid might prove to be of value to a cask user in an as-yet-undefined mode of operation.

2. The original design of the basket used aluminum throughout due to its structural and thermal properties combined with light weight. The drop tests were carried out with this design and demonstrated the high margin of safety of the basket. One test subjected the basket to the equivalent of 200g with only minor deformation. Nevertheless, the NRC rejected the use of aluminum for structural reasons citing the lack of inclusion of the aluminum in the ASME Code. As a result, steel disks were interspersed with aluminum disks with the provision that all structural requirements be met by the steel disks. The heat transfer properties of the aluminum have been retained as described earlier.

3. Other detailed basket design features were of concern to the NRC and required design features to be added. Among these were extra drain holes in the basket to assure no "hide-out" of water during draining notwithstanding the basket configuration is a very "open" design. In addition, the tie rods stretching the full length of the basket and aligning all of the disks were required to have tight tolerances on their spacer locations. This was to assure precise orientation of the disks in accordance with analytic evaluation assumptions in structural evaluations.

4. Borated aluminum tubes were used in the original design bases of the basket consistent with the weight benefit, heat transfer benefit and criticality features needed, but with recognition of the lack of any structural capacity assigned to the tubes consistent with NRC preferences regarding aluminum. However, the rejection of the use of aluminum by the NRC, particularly borated in this case, caused a redesign to include the present approved configuration using Boral completely sealed in stainless steel sheets but with the main structural features of the basket in the steel support disks.

5. Features related to thermal conditions within the package caused additional NRC requirements. For one, it was required that basket expansion in the early heat up phase following fuel loading and drying should not cause impingement on the slower responding cask body during the period up to thermal equilibrium. This resulted in very detailed analyses and subsequent tolerances requirements with due consideration of different materials expansion rates. Specifying a large gap between basket and body did not suffice since this would impair heat transfer. Similarly, thermal expansion of the NS4FR neutron shielding material during the period up to thermal equilibrium was registered as a concern of potentially imposing loads on the outer containment shell and the outer skin both of which form the cavity for the NS4FR. Special expansion foam sheets were added to the cavity inner surface to absorb any such loads if they occur.

6. A final conservatism required by NRC that relates to thermal performance, is the execution of a full-scale thermal test with full-design basis kW load using heaters in the cask. A measurement of outer-skin temperatures compared to those

predicted is required. While a type of thermal test has been required for other designs in the past, generally the as-loaded fuel source has been sufficient. In this case, a full-scale test on every STC unit fabricated is required prior to release of the unit for commercial use.

7. Key welds of the cask containment are subjected to full nondestructive examination (NDE) testing according to ASME requirements. However, in addition, other noncontainment welds are required to have full volumetric testing. This is beyond code requirements. Also, a maximum thickness of 6mm on each pass was imposed. These requirements are clearly beyond any previous experience NAC has had.

3.0 Use Of The NAC-STC In Canister Mode

It has been NAC's experience that once NRC approves a cask for transport of design-basis spent fuel contents, it is relatively easy to request NRC approval of a variety of contents that are different, but do not impose additional or new conditions on the cask, i.e., criticality, heat load, weight, shielding are all within limits. In many cases, these differing contents have included canistered fuel. Worldwide, NAC has shipped several hundred packages in a canistered configuration. A current example of a new configuration just recently licensed in December 1995 is a 25-element high-enriched metallic-test reactor (MTR) fuel contents in the NAC legal-weight truck cask originally licensed for one PWR or two BWR assemblies. In this case, the new configuration is to allow movement of the Georgia Institute of Technology Research Reactor fuel to Savannah River. This movement actually occurred in February 1996.

In a similar manner, the NAC-STC can be utilized for some fuel types in a canistered mode. The early "senior citizen" plant group of which Yankee Rowe is an example, have a fuel type that can be accommodated in the NAC-STC in a canister. Figures 7 and 8 show a slightly modified tube and disk design basket that can accommodate 37 fuel elements of this type. This basket uses the same key dimensional features as the STC such as 13mm thick steel disks (although fewer are needed) and the "web" thickness between fuel tubes. In this way, the same basket thermal and structural performance is achieved; the fuel radiation source term, the thermal fuel loading, and the contents weight are all within the STC license basis.

Enclosing the basket in a 16mm thick wall welded canister is an easy addition that adds to the overall safety of this utilization of the STC. In this way, a transportable canister and basket is achieved. Since it is compatible with an already licensed transport cask, this canister may be transported as soon as there is a designated interim storage site. All that remains then is for a concrete storage-type cask, as shown in figure 9, to be sized compatible with the transportable canister to have a complete canistered type or MPC[2] system suitable for this class of plants.

---

[2] The USDOE has defined a canister system as a multi-purpose canister system (or MPC) if it can be stored and transported with spent fuel in the canister.

The next logical step in the development of the NAC transportable storage technology is the design of a universal MPC system hereafter referred to as the Universal Multipurpose Canister System (MPC) System™ (UMS™). A description of this system is shown in figures 10. In this case, the same key features of the components are retained, but extended to cover all of the fuel types and their dimensions. The canister in this case has a 1.73cm outside diameter by 6.73cm in length and continues to use the proven tube and disk arrangement. The transport cask (or overpack as it is sometimes called) is about 7% longer than the NAC dual-purpose cask, i.e., 5287mm compared to 4090mm. Its diameter is correspondingly reduced 7% from 249mm to 231mm in order to keep the overall weight at 125 tons in the fully loaded condition ready for transport. With these modifications, the UMS™ has a capacity of 24 PWR assemblies or 56 BWR assemblies. Correspondingly, the concrete cask or overpack dimensions are compatible with the canister.

The key components of the UMS™ are a logical and conservative step forward from the existing licensed technology. As a result, the preparation of the required storage safety analysis report (SAR) and the transport safety analysis report (TSAR) will follow the same pattern laid out in the corresponding NAC-STC SARS. The bases of the analyses, including the modeling methods, the conservatism, the design criteria and the design features are all maintained. Testing will be repeated to confirm transport margins as in the past.

The project schedule, started in January 1996, shows a transport SAR submitted to NRC in September, 1996 followed by a storage SAR in October of the same year. Early discussions have already taken place with the NRC in November 1995, and consequently a storage C of C in late 1997 and a transport C of C in early 1998 are planned. This will allow initial fabrication of the storage components of the UMS™ to be ready for use by mid 1998. The transportation portion of the UMS™ could begin fabrication shortly thereafter if there were a centralized interim storage site available. For these reasons, the UMS™ will be accomplished faster than other systems.

The UMS™ project will be accomplished by a team of respected design and fabricator companies located throughout the world. In addition, worldwide utility interest is resulting in cost sharing of project expenses. At the present time, the UMS™ design, testing and fabrication team is comprised of:

NAC International, Atlanta
TransNuclear, New York
ABB Combustion, Connecticut
ENSA, Spain
Hitachi Zosen, Japan
SAMSUNG, South Korea

In addition, utility companies and others who have agreed or are considering sharing the costs of the project are:

Arizona Public Service Company, Arizona
Virginia Power Company, Virginia
ENRESA, Spain
Tokyo Electric Power Company, Japan
Kansai Electric Power Company, Japan
ITOCHU Trading Company, Japan

As a result of this level of participation, the UMS™ is clearly an international project. With this level of participation and interest, the resources to accomplish the program on schedule is assured. Some of the team design members had already planned a similar project scope sharing as a consequence of bidding on the DOE-MPC program. Many of the members of the UMS™ team are familiar with the NAC technology; some are already using NAC systems, some have already fabricated the NAC systems, and some have also designed "sister" models for their own applications and are in the process of receiving their NRC equivalent approvals.

As can be seen, the UMS™ project is totally privately supported. The projected cost is a small fraction of that estimated for the DOE-MPC project covering the same scope. Since there is cost sharing by the UMS™ team members, the cost to any one member is substantially lower than shouldering the total project.

Internationally speaking, the cost to any one nation's industry is also less. After NRC and IAEA certificates for the UMS™ are obtained, other international team members will prepare safety reports in their own language, adding any special requirements specified by their regulatory agencies, as appropriate prior to approval and use in those countries.

It is apparent that several of the UMS™ team members are fabricators. Their contributions in the early phases of the project are to inject fabrication technology into the details of the design and assure optimizing the manufacturing phase of the UMS™. In the later stages of the UMS™ project, clearly these same fabricators will have an early lead in their industry to service their local markets. The added benefits of having a worldwide source of qualified fabricators will improve the competitiveness of UMS™ systems delivered to any end user, once again contributing to the project objectives.

In conclusion, the UMS™ project will fill the needs of the U.S. utilities who are generating spent fuel. The design uses the same features and incorporates the same materials as the NAC-STC, which is already approved by the NRC. Thus the NRC approval should be accomplished in a timely fashion consistent with the growing number of U.S. utilities running out of storage space. The team assembled to execute the UMS™ design and licensing have accumulated more experience than any other organization in the U.S. industry. It is believed that the UMS™ program will fill the void left by the DOE decision to unfund their MPC program. An early implementation of an MPC system, such as for the Yankee type of plants, is already under way. The participation of the worldwide highly qualified partners will ensure a successful quality product at a competitive price.

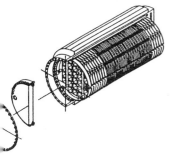

Figure 1—NAC Transportation Overpack

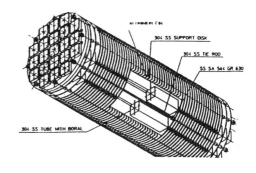

ALUMINUM FIN
304 SS SUPPORT DISK
304 SS TIE ROD
SS SA 564 GR 630
304 SS TUBE WITH BORAL

Figure 2—Fuel Type Envelope, NAC-STC Design Basis

Figure 3—NAC Fuel Basket

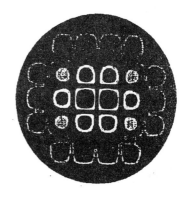

Figure 4—NAC Impact Limiter Technology

Figure 5—NAC Fuel Basket Thermal Performance

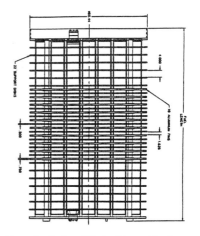

Figure 6—Cask and Basket Thermal Performance

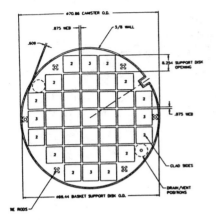

Figure 7—NAC-MPC(Y) Basket

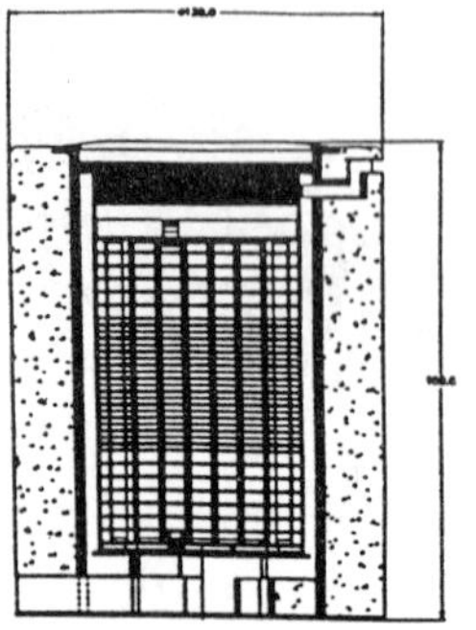

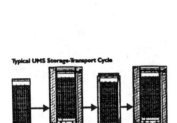

Figure 8—NAC-MPC(Y) Canister

Figure 9—NAC-MPC(Y) Concrete Overpack

Manufacturer	Type Included	Comment
Westinghouse	14 x 14	inside envelope
	15 x 15	Criticality Source
	17 x 17	Fuel Source Term
		(Shielding)
Combustion	14 x 14	inside envelope
Engineering	15 x 15	inside envelope
	17 x 17	inside envelope
Babcock and	15 x 15	inside envelope
Wilcox	17 x 17	inside envelope
Total Heat Load	22.1 kW	

40,000 MWD/MTU and 6.5-year cool time
or
45,000 MWD/MTU and 10-year cool time

NAC-STC able to handle about 75% of all PWR Fuel (length

Figure 10

The objectives and benefits of GEC Alsthom modular vault dry store technology for interim dry storage

C CARTER BTech, CEng, MIMechE
GEC Alsthom Engineering Systems Limited, Leicester, UK

SYNOPSIS

Dry Storage of spent nuclear fuel is now a well established and accepted part of the nuclear fuel cycle. There are several dry fuel storage technologies in the commercial market that are available to a nuclear utility. This paper describes one such leading dry storage that is known as the Modular Vault Dry Store, or "MVDS". The MVDS technology is a concrete vault storage system that has an operational background of over twenty five years. The MVDS is a flexible design approach that is suitable for the storage of all types of spent nuclear fuel or high level waste streams.

THE OBJECTIVE OF THE MODULAR VAULT DRY STORE SYSTEM

The MVDS, as illustrated overpage on Figure 1, is designed to provide an interim storage facility for keeping spent nuclear fuel in a safe, dry, non-degrading state. Fuel that is removed from a dry storage facility after its term of residence should be discharged in a comparable form to when it was first loaded, so that all future disposal options are maintained. The dry fuel store has to provide engineered features that ensure, by design, that the condition of the fuel is maintained without degradation. The principle conditions needed to ensure correct dry storage of spent nuclear fuel are: proper dryness of the fuel, inert gas cover, high integrity sealed containment and efficient cooling of the fuel in store. The length of the term "interim" can mean anything from 5 years to 125 years in store, but is very much conditioned by when the next alternative step is available within the fuel cycle. Dry fuel storage facilities are not meant to become an indefinite fuel repository nor to provide a final solution to close the fuel cycle.

Why is interim storage of spent nuclear fuel required? Currently there are two main final disposal solutions available for spent nuclear fuel: either reprocessing and vitrification or direct disposal. If direct disposal or reprocessing facilities are not available to a utility, or if reprocessing is politically unacceptable or too expensive, then in such circumstances the continued operation of the nuclear power plants will require facilities for extended, ie interim, storage until the issues relating to the final disposal solution have been resolved.

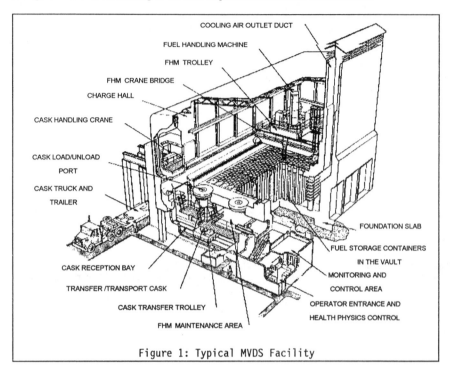

Figure 1: Typical MVDS Facility

The success of dry storage within the nuclear fuel cycle can be attributed to several advantages over the wet storage option:

- Dry storage of spent nuclear fuel maintains the condition of the fuel without significant degradation.

- A wider range of fuel types can be stored for interim periods in dry conditions than can be stored wet.

- Dry storage safety systems for spent nuclear fuel are passive and therefore require very little operational maintenance and produce no operational wastes during operation.

- Dose uptake to the operators is low during loading and operation.

- Dry fuel storage systems are cheaper to construct and have lower operating costs compared with equivalent sized new wet storage systems.

Existing wet storage systems will continue to be used for interim storage of fuel, especially for fuel awaiting reprocessing. Dry fuel storage is, however, almost universally now accepted as the industry standard for new facilities for fuel storage.

THE DEVELOPMENT OF THE MVDS TECHNOLOGY

The MVDS technology was developed directly from the original Magnox fuel dry storage facilities at the Wylfa power station in Wales (UK). The Wylfa dry stores were originally commissioned in 1969 and the MVDS can, therefore, rightly claim to be the only dry storage technology that has an operational and technological background of a quarter of a century. The original design features of the Wylfa dry stores that have been carried over into the MVDS design basis are:

- The single carbon steel containment boundary connected to a source of the gaseous storage environment.

- The natural cooling air flow across the outside of the tubes arranged to move horizontally across an array of vertical tubes.

- Concrete radiation shielding which also helps in delineating the air flow by creating inlet and outlet ducts.

Figure 2: Fort St Vrain MVDS

The MVDS technology has been extensively developed over the last twenty five years so that it can be applied to the storage of a range of alternative fuel types. The design and development of the Modular Vault Dry Store have been carried out by GEC ALSTHOM Engineering Systems Ltd in the UK. The MVDS was jointly developed for the USA market together with the US licensee Foster Wheeler Energy Corporation of Perryville, NJ. The MVDS has had a USNRC Topical Report approved status since 1988. The Topical Report covers the interim storage of Light Water Reactor fuels, including both PWR and BWR, at any reactor site in the USA. A representative "Topical Report" style MVDS facility is shown in Figure 1.

The first contract for an MVDS to be constructed in the United States was placed by the Public Service Company of Colorado in August 1989, to be built at their Fort St Vrain high Temperature Gas Reactor site. The Fort St Vrain MVDS is shown on Figure 2. The complete Fort St Vrain MVDS project for the design, licensing and construction of the six vault facility took 25 months. The Fort St Vrain MVDS was handed over to the client in September 1991 and has been loaded with 1,488 HTGR fuel blocks.

The MVDS technology is presently being applied to the storage of VVER fuel in Hungary where the Safety Analysis Report has been approved and construction is proceeding. The Paks MVDS is designed to hold 5,000 VVER 440 spent fuel assemblies and can be extended to hold up to 15,000 VVER 440 spent fuel assemblies. The Paks MVDS will receive fuel in 1996.

DESCRIPTION OF THE MVDS SYSTEM

The MVDS as shown on Figures 1 and 2 was designed and developed as a spent fuel management system. There are two basic variations to the MVDS design and these differ primarily in the fuel handling philosophy. Both systems have been licensed and can be adapted for the storage of a wide range of fuel types and nuclear materials.

In one system the fuel is handled as a bare fuel assembly within the transfer cask and Fuel Handling Machine through the fuel transfer route. These bare fuel assemblies are subsequently placed into sealed storage tubes at the vault. The Paks MVDS is based on this design. In the alternate system the fuel is placed into the transfer cask within sealed containers at the reactor pool. The fuel assemblies remain inside the sealed containers as they pass through the route in the cask and Fuel Handling Machine. The containers are placed into the vault to provide the storage location for the fuel. The Fort St Vrain MVDS is based on this design. The bare assembly transfer system allows fuel to be loaded at the reactor pool into existing transportation casks, either wet or dry, and minimizes any operational changes needed at the reactor pool. The containerized fuel transfer system involves some operational changes at the reactor pool associated with the sealing of the container but provides the subsequent advantage of a contamination free transfer route to the MVDS.

The MVDS storage facility, as shown on Figure 1, consists of three main systems:

i)   The Cask Reception Bay, where spent fuel is received (and despatched at the end of life).

ii)   The Storage Vault Modules, where spent fuel is stored.

iii)   The Fuel Handling Machine, which raises and transfers irradiated fuel assemblies, or containers, from the cask in the Reception Bay to the storage position in the Storage Vault.

Following the cross-site, or cross-country, transfer of the fuel to the MVDS, the cask is placed in position in the Reception Bay. After venting the cask, the lid bolts are released to allow the removal of the lid. The lid is removed remotely and the cask is then moved to a position where it is seismically clamped to allow access to the fuel assemblies (or containers) by the Fuel Handling Machine.

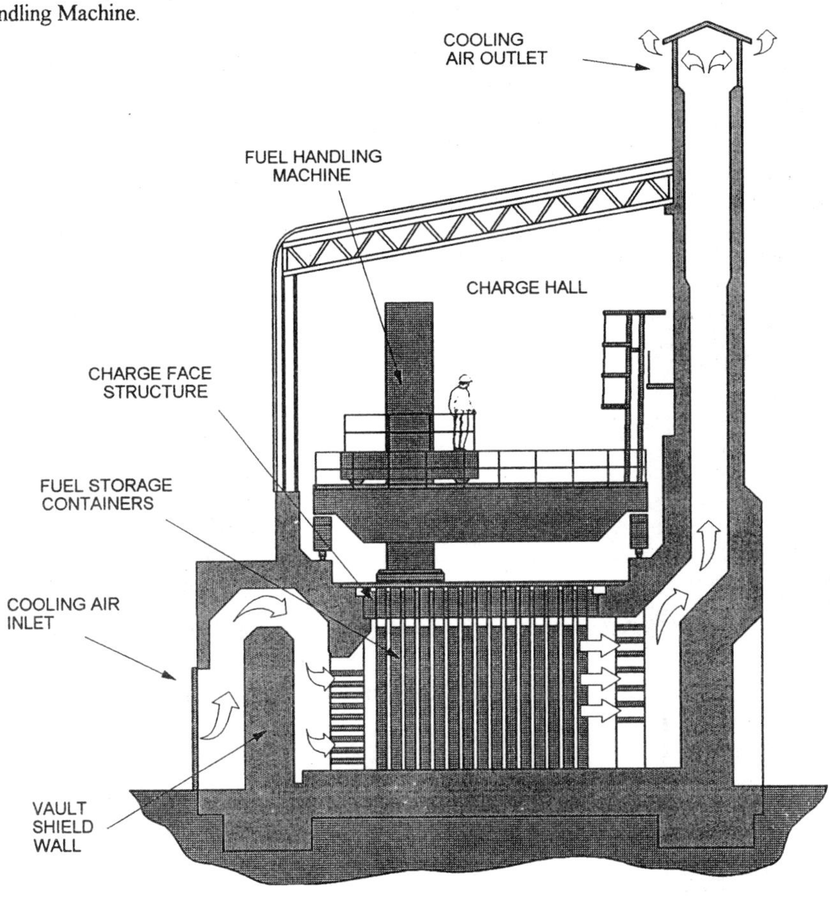

Figure 3: Cross Section Through MVDS

The Fuel Handling Machine is designed as a heavy shielded structure to allow the safe transfer of the fuel assemblies or containers from the Reception Bay to the storage location at the Vault. The Fuel Handling Machine design can be based on either a bridge-mounted machine or a free-standing machine moved around by an overhead travelling crane. The Paks MVDS uses a bridge mounted Fuel Handling Machine and the Fort St Vrain MVDS uses a free standing, crane carried Fuel Handling Machine.

The MVDS cooling system is thermally very efficient and designed to give low fuel storage temperatures. The principle of operation of the MVDS is illustrated on Figure 3. This figure shows the passive cooling air flow system, the vertical position of the fuel within the vault and the concrete shield walls that surround the fuel to form the vault structure.

The fuel assemblies are stored within the vault in either Fuel Storage Containers or Fuel Storage Tubes. The difference between a Fuel Storage Container and a Fuel Storage Tube is that the Fuel Storage Container has a bolted closure lid sealed by metallic seals, while the Fuel Storage Tube has a shielded closure plug sealed by elastomeric seals that can be removed by the Fuel Handling Machine. The two variants are shown in Figure 4. The Fuel Storage Container/Tube can each hold either a single fuel assembly or multiple assemblies. The Fuel Storage Container/Tube are located in the vault by the Charge Face Structure at their upper end, and in a Support Stool at the lower end. The Charge Face Structure provides shielding from the fuel to allow access over the tube top. It is the design of the Charge Face Structure that permits the monitoring and inspection of the fuel during storage, if required.

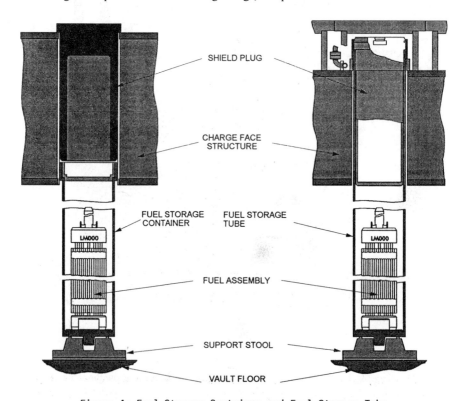

Figure 4: Fuel Storage Container and Fuel Storage Tube

The ability to monitor the fuel containment is an integral part of the fuel management function. The MVDS design permits the containment boundary and storage environment of the Storage Container/Tube to be monitored throughout the life of the store. Each fuel Storage Container/Tube can be permanently, or periodically, connected to an inert cover gas supply and monitoring system. The monitoring system enables leakage to be detected from the fuel containment boundary by monitoring gas movement (flow) or system pressure.

At the end of the storage life when the MVDS has to be defuelled, the operations are the reverse of the loading process. Because the future requirements for road and rail transport cannot be accurately predicted today, it cannot be guaranteed that the same transfer cask used for loading the MVDS will be suitable for the unloading operations. The flexibility of the MVDS Cask Reception Bay enables it to accept a wide range of cask options and in the event that another type of cask has to be handled, then a design modification to the Reception Bay would be relatively simple. Fuel is discharged off-site directly from the MVDS; there is no requirement for the reactor pool to be available for this operation.

BENEFITS OF THE MODULAR VAULT DRY STORE

The MVDS is designed to provide a safe storage environment for spent nuclear fuel for an interim period. Spent fuel has to be managed in a dry (or wet) storage system to ensure that the recovery of the fuel can be achieved with minimum cost and exposure. The design of a dry fuel store has a big impact on the ability to recover and transport spent fuel in the future. The MVDS as a candidate technology for the interim dry storage option offers the following fuel management attributes and benefits:

i) Independence of the System

The MVDS is a totally independent stand alone system. All fuel handling operations, both for receipt and discharge, are accomplished using the equipment provided at the MVDS. The fuel transfer operations take place directly from the fuel transfer flask into the Fuel Handling Machine and into the storage position.

ii) Passive Self Regulating Cooling System

The fuel assemblies are cooled by a passive self-regulating cooling system that induces buoyancy-driven ambient air to flow across the exterior of the individual Storage Tubes. Not only does this provide a totally passive cooling system, the storage temperatures for the fuel assemblies provide high margins against the allowable temperatures. The MVDS is capable of storing all types of fuel, including LWR, HTR, AGR, Magnox, consolidated fuel and vitrified waste.

iii) Sub Criticality

In the MVDS the fuel assemblies are stored within Storage Containers/Tubes that provide a fixed storage geometry and ensures that criticality is not possible under all normal and fault conditions, including partial or total flooding. The design takes no credit for fuel burn-up. The inherent safety of the system is therefore easier to demonstrate and allows significant operational flexibility when having to store fuel with varying radiological characteristics and histories.

iv) Diversity of Safety

The shielding, cooling and confinement functions for the fuel assemblies are provided by three diverse systems. The massive concrete walls provide the biological shielding, the Storage Containers/Tubes provide the confinement and the passive buoyancy driven air provides the cooling. The Fuel Handling Machine is a heavily shielded transfer device which minimizes personnel dose uptake and provides a passive cooling system during fuel loading and unloading operations.

v) Fuel Storage Condition

During the period of storage in the MVDS the storage environment is designed to prevent degradation of the fuel assemblies. A high purity inert gas is selected as the storage environment. The MVDS design permits the Storage Containers/Tubes to be connected to a high integrity source of inert gas and a monitoring system which allows the condition of the cover gas and the integrity of the storage container to be checked on a continuous basis. The method of loading the fuel assemblies into the MVDS can provide the opportunity to check the condition of the fuel assemblies as received or to remove them for periodic inspection.

vi) Monitoring

The condition of the fuel assemblies over long periods in store can be monitored in the MVDS system. The ability to monitor the fuel condition is an inherent part of the waste management function.

vii) Loading Flexibility

The MVDS was designed as a modular system and thereby provides the ability to add storage capacity on a phased modular basis. The MVDS is designed to allow the loading operations to take place continuously through all weathers, seasons and during the day or night by the addition of a roof over the Charge Hall. The MVDS Charge Hall is weather proof, can be heated and illuminated, and is designed to withstand severe wind and snow loadings, thereby allowing the highest operational availability.

viii) I.A.E.A. Safeguards

The nuclear security features of a fuel storage system must be commensurate with nature of the materials being stored and the public and political demands for their safekeeping. The MVDS is a vault system, and as such sets the standard for nuclear safeguards and security considerations. Fuel is received, prepared and placed into storage all within one building. Surveillance of the incoming and outgoing fuel handling facilities is therefore simplified and risks of diversion are reduced. The single receipt and storage building for the MVDS minimises the size of the site and the length of the boundary fence and therefore enhances security at the lower cost. In addition, the fuel Storage Containers/Tubes can be accessed by IAEA Inspectors for direct visual inspection with minimal operational difficulty and cost. It is possible to provide TV viewing facilities so that the reference number on the top of each fuel Storage Container/Tube can be viewed using the fuel handling equipment without moving the fuel from its storage location.

FUTURE DEVELOPMENTS OF THE MVDS TECHNOLOGY

Applications have been already been found for Magnox, High Temperature Reactor, and Light Water Reactor VVER fuels, and non-site specific licence approval has been gained for US Boiling Water and Pressurised Water Reactor fuels. This diversity of fuel type with their widely differing characteristics demonstrates the adaptability of the MVDS concept. The number of national licensing authorities that have now examined site specific designs also provides confidence that the planning timescale for the licensing procedures can be assured.

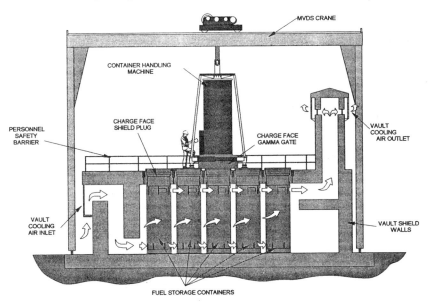

Figure 5: Section Through MPC Style MVDS

The latest development to the MVDS has been to integrate and rationalize the design for use with the Multi Purpose Container technology, or MPC, as shown above on Figure 5. MPC is a generic term that refers to a multi-position, multi-purpose container for storing spent nuclear fuel. The MPC concept began in the USA several years ago. The principle of the MPC is to provide a sealed-for-life storage container into which the fuel is placed, and for the fuel never to have to leave that container before it reaches its ultimate disposal location. The MPC design can embrace the principles of storage only, or both storage and transportation. There are several designs of MPC that are needed to meet the different fuel types, sizes and radiological characteristics.

One change from the original design basis of the MVDS is that the MPC is used to store a large number of fuel assemblies in one container, typically 24 PWR spent fuel assemblies or 44 BWR spent fuel assemblies. The diameter of the MPC is about 1600 mm.

The opportunity has been taken to review the design basis of the MVDS and to remove some of those features that do not affect the fundamental safety approval of the facility. The profile of the MVDS building has been lowered to improve the visual impact of the installation, this has been achieved by removing the covered roof area and lowering the height of the outlet duct. A single vault can accept fifteen MPCs, giving a total vault heat dissipation capability of approximately 350kW. The MPC Handling Machine, together with its overhead crane is illustrated below on Figure 6.

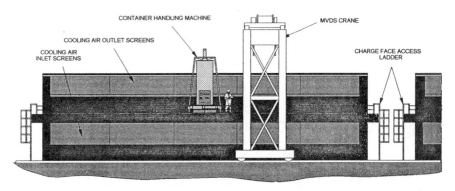

Figure 6: Side Elevation View on MPC Style MVDS

The use of the MPC with the MVDS technology produces a fuel storage solution that is compact, simple and cost effective. The use of a transportable MPC provides an opportunity to ease the future recovery and off-site transportation of the spent fuel. The transportable MPC never has to be opened prior to transport, the complete container is simply lifted out of the vault using the Container Handling Machine and crane, and lowered back into a road transport qualified cask. Handling multiple fuel assemblies in one container means that the transfer times are quick, manning levels are low, and therefore dose uptake during these operations is very much in accordance with the ALARA principle.

CONCLUSION

The back end of the nuclear fuel cycle urgently needs to be completed to prevent significant accumulations of spent fuel. Direct disposal of fuel or reprocessing are currently the two main options for completing the nuclear fuel cycle.

If reprocessing of fuel is either not available or not acceptable to a utility, then the present lack of final geological repositories around the world will cause nuclear generating utilities to have to store fuel on site for extended periods of time. The MVDS is designed to provide an interim solution to that problem. The MVDS technology is mature, proven, licensed and suitable for dry storage of a wide range of spent fuel types or high level wastes.

MACSTOR dry spent fuel storage system

E PARÉ BASc, MemCanNucSoc, MemOrdEngQueb and **P PATTANTYUS** MemOrdEngQueb
AECL, Montreal, Canada

SYNOPSIS

A world pioneer in many aspects of nuclear technology, **AECL** (Atomic Energy of Canada) Limited) has also assumed leadership in the area of dry storage of spent fuel. The Canadian Crown Corporation first started to look into dry storage for the management of its spent nuclear fuel in the early 1970's. After developing silo-like structures called concrete canisters for the storage of its research reactor enriched uranium fuel, AECL went on to perfect that technology for spent CANDU (Canadian Deuterium Uranium Reactor) natural uranium fuel.

From the early AECL-designed unventilated concrete canisters to the advanced **MACSTOR** concept - **M**odular **A**ir-**C**ooled **C**anister **Stor**age - now available for LWR fuel dry, storage is proving to be safe, economical, practical and, most of all, well accepted by the general public. AECL's experience with different fuels and circumstances has been conclusive.

1. INTRODUCTION

Dry spent fuel storage, especially for Nuclear Power Reactors, is increasingly gaining acceptance worldwide. This storage method has several advantages over wet storage. Because of its passive functional nature, operation and maintenance costs are low and no complex facilities are needed. Capacity increase is incremental and adjustable. Dry storage facilities can be designed in a way that the fuel can be shipped offsite without transiting back through the pool.

Spent fuel transportation relied on steel casks since the early stages of nuclear power development. This resulted in steel casks also being used for dry storage. The drawbacks of steel storage casks are cost and performance limitations. The heat rejection through conduction limits their capacity because the fuel clad temperature has to be maintained at safe levels during the storage period.

An alternate storage technology to steel casks was pioneered by AECL in the early 1970's. CANDU Power Reactor and Research Reactor spent fuels have been experimented with at AECL's Whiteshell facility. Dry spent fuel storage was required at Whiteshell because no pool facilities were available. Concrete was selected and its suitability demonstrated to safely and economically store spent fuel. As a next step, in 1990, AECL together with Transnuclear Inc. (TNI), a U.S. based company, began to develop an air cooled concrete storage system called MACSTOR.

Derived from AECL's successful vertical loading, concrete silo program for storing CANDU nuclear spent fuel, MACSTOR was developed for light water reactor spent fuel and was subjected to full scale thermal testing at AECL's Whiteshell Laboratories. This series of tests demonstrated conclusively that the massive air currents inherent in a MACSTOR design removed the required amount of heat to maintain fuel and concrete temperatures at acceptably low values.

The MACSTOR Module is a monolithic, shielded concrete vault structure that can accommodate up to 20 spent fuel canisters. Each canister typically holds up to 21 PWR or 44 BWR spent fuel assemblies with a nominal fuel burn up rate of 40,000 MWD/MTU and a 7 year minimum cooling period. The structure is passively cooled by natural convection through an array of inlet and outlet gratings and galleries serving a central plenum where the canisters are vertically located.

The MACSTOR system includes the storage module(s) and canisters, an overhead gantry system for cask handling, a transfer cask for moving fuel from wet to dry storage and a cask transporter. The canister and transfer cask designs are based on TNI's transport cask designs adapted to requirements for on-site spent fuel storage.

The MACSTOR system can economically address a wide range of storage capacity requirements. The modular concept allows for flexibility in determining each module's capacity. Starting with 8 canisters, the capacity can be increased by increments of 4 up to 20 canisters.

Various containment types can be accommodated by the MACSTOR system. The MACSTOR Module fitted with storage cylinders provides additional containment for each canister in case of lesser thermal performance requirements. The single containment method is applied when higher heat release has to be dealt with.

The MACSTOR system is also flexible in accommodating the various spent fuel types (besides the common PWR & BWR assemblies) from such reactors as VVER-440, VVER-1000 & RBMK 1500.

Full system compatibility can also be achieved with the U.S. Department of Energy's Multi Purpose Canister (MPC) concept.

Hydro-Québec was the first utility to implement the MACSTOR system to store spent fuel from its CANDU type power reactor at the Gentilly 2 station. The AECL designed storage system was licensed by the AECB (Atomic Energy Control Board), the Canadian Regulator. The first storage module was built during the summer of 1995 and fuel loading started in September 1995.

2. AECL EXPERIENCE AND DEVELOPMENT OF VERTICAL MODULAR DRY STORAGE FACILITIES FOR SPENT FUEL

AECL is Canada's nuclear organization, established by the Canadian government in 1952 to develop peaceful applications of nuclear energy. To carry out this mandate, AECL conducts research and development for nuclear-related products and services; designs and markets CANDU nuclear power stations; provides engineering services to electrical utilities in Canada and internationally; and markets a wide variety of products and services.

AECL's experience with spent fuel management goes back over 40 years in Canada. The Canadian wet storage technology was initially developed by AECL's research facilities, at the Chalk River (CRL) and Whiteshell (WL) Laboratories. In 1974, WL began development and demonstration of a dry storage system, based on a "concrete canister" (CC), (which is an unventilated concrete storage cask), as a possible alternative to the storage of spent CANDU fuel in storage pools. The CC is a thick walled concrete monolith, containing baskets of fuel in the dry state.

The typical CANDU reactor fuel bundle, containing 37 natural uranium dioxide elements, weighs about 22 Kg and is 10 cm in diameter and 50 cm long. Due to its relatively small size, freedom from criticality hazards in light water and relatively low burnup characteristics (a typical average burnup being 7800 MWd/MgU), CANDU fuel is amenable to simple and basic storage systems.

Four CCs were designed and constructed at WL. Two containing electric heaters were subjected to heat loads of 2.5 times the design, ramp heat-load cycling, and to weathering tests with irradiated fuel. The collected data were used to verify the analytical tools for prediction of effectiveness of heat transfer and for radiation shielding and to verify the design of concrete structure.

The imposed design limits were:

- 250°C for hottest fuel element, in an inert gas atmosphere,

- 0.2 mm crack width on the external concrete surfaces,

- 100 micro Sievert/h exposure rate on the external concrete surfaces.

The fuel handling methodology consists of loading round stainless steel baskets underwater with fuel bundles. Following placement of the basket cover, the basket is lifted into an aboveground shielded work station. Drying of the fuel bundles is followed by remotely welding the basket cover. Each basket is then hoisted into a transfer cask, which provides adequate shielding. The cask is moved to the storage site and lifted over the CC's loading platform. Finally the basket is lowered into a steel lined cavity.

By 1987, CCs were being used to safely and economically store all spent fuel generated by AECL's decommissioned prototype reactors.

In 1989, New Brunswick Power selected AECL's dry storage option to extend the on-site spent fuel storage capacity of its CANDU 6 Nuclear Generating Station (NGS) at Point Lepreau.

The CCs are being built at the site and are vertically loaded with sealed storage baskets containing CANDU spent fuel bundles.

In 1991, two CANDU 6 stations (Wolsong 1 in Korea and Gentilly 2 in Canada) adopted AECL's technology for their dry spent fuel storage needs.

Dry storage in CCs evolved progressively, starting with small capacity. The capacity of the first CCs, containing 6 small baskets used at WL, was increased to hold 9 larger baskets at Point Lepreau.

Further enhancement of AECL's concrete based dry spent fuel storage technology started in 1989 through the introduction of a vault type storage concept. In this approach, several CC's are consolidated into a single, large concrete storage module. While the spent fuel in the CC's is cooled solely by heat conduction, the new concept incorporated passive convection cooling.

The storage module approach offers several benefits. While significant cost and storage space savings can be achieved, higher thermal loads can also be accomodated.

The first MACSTOR type, air cooled concrete monolith structure (designated CANSTOR) for storage of CANDU spent fuel was constructed at Gentilly 2 NGS in 1995. Please refer to Figure 1.

AECL's experience with dry spent fuel storage is summarized in Table 1.

3.   THE MACSTOR SYSTEM

AECL has considerable experience in the storage of CANDU spent fuel in CCs. This concrete technology can be further developed for use in LWR fuel storage and economically be made attractive (compared to metal cask storage) to the potential users.

While AECL's concrete technology can be developed for LWR fuel storage, the methodology used for handling CANDU fuel is not readily applicable to LWR fuel. This is due to the fact that LWR fuel assemblies are very long compared to length of CANDU fuel. Therefore, it is necessary to adopt the technology that is currently practised to remove LWR fuel from wet storage. A strategic decision was made to join forces with an experienced LWR fuel handling company to develop a hybrid system which could combine the operational economies of metal cask technology with the capital economies of concrete technology.

AECL thus teamed up with TNI, a well established vendor of Nuclear Fuel Handling Services, to jointly develop, market and execute the MACSTOR system.

TNI supplies equipment to remove the fuel from the pool in a metal storage canister housed in a cask and a transporter vehicle to bring the cask to the storage site. AECL's loading system then takes the metal canister form the transporter vehicle and place it in the concrete vault. TNI provides the design and procurement of the Transfer Cask, the Transfer Vehicle and the Storage Canister which will be stored inside the concrete vault. AECL provides the design and the procurement of the hoist, the "canister transfer system", and the design and construction of the MACSTOR module.

The MACSTOR Module

The modular concept allows for flexibility in determining each module's capacity. Starting with 8 canisters, the capacity can be increased by increments of 4 up to 20 canisters.

The MACSTOR Module, shown in Figure 2, is a rectangular, reinforced concrete structure, 27 feet in width, 67 feet in length and 21 feet in height. It stores 20 canisters. Each canister is vertically supported by an impact-limiting structure affixed to the floor of the Module and a collar embedded in the top slab of the Module.

There are five inlet air ports located in each of the two longitudinal walls, near the base of the structure, and six outlet air ports in each of the same walls, located just below the top slab. The air inlets are located above ground to protect the Module from water ingress and also from blockage due to snow accumulation. Protective weather covers are in place over each of the canister openings on top of the Module to prevent the possible ingress of rain, snow or foreign objects. These weather covers are anchored to the top slab of the Module.

The Module design meets the requirements of US 10CFR 72, "Licensing Requirements for the Independent Storage of Spent Nuclear Fuel and High Level Radioactive Waste".

There are three salient technical aspects of the design of a concrete monolith for the storage of spent LWR fuel. These are: the heat rejection capability, the shielding adequacy and the structural soundness. Of these, only the heat rejection capability represented an area of potential technical uncertainty to AECL at the early design stage. The structural adequacy and shielding adequacy of the MACSTOR design were of no concern because of AECL's very extensive concrete canister experience in these areas. The CANSTOR/MACSTOR thermal tests performed (at Whiteshell in 1990) served to remove the technical uncertainty that existed with respect to the capability of the Module to reject the heat generated by LWR fuel.

The MACSTOR Canisters

Various types of containment can be accommodated by the MACSTOR system as a function of the thermal performance requirements.

The fully cooled and sealed canister can be placed inside a storage cylinder which is permanently installed in the MACSTOR Module. By sealing the storage cylinder following canister loading, a double- containment feature is achieved. This method is used when the heat release or the canister capacity is relatively low. For higher thermal performances, the canister alone is providing a single containment. The double containment approach allows for designating either the canister or the storage cylinder as containment boundary for licensing proposes. The choice depends on the boundary which is better qualified for the containment function, i.e. closure welds can be examined and the boundary's leak tightness can be verified.

The double containment approach also facilitates monitoring of the canister leak tightness. By analyzing samples taken from the storage cylinder containing the canister, the presence of Helium can be detected. Defective canisters can still be safely stored by pressurizing the storage cylinder. Table 2 describes the characteristics of the canister used in various containment types.

The canister, containing the fuel basket, consists of a cylindrical steel shell with an integrally welded head at the bottom. The fuel basket is an assembly of stainless steel cells joined by welding with an aluminum thermal conductor and a neutron poison plate between the cell boxes which are fusion welded together at many locations along their length, forming a strong honey-comb like structure which provides compartments for the fuel assemblies.

The canister closure system (which also provides shielding at the top) can be either of the welded or of the flanged type. The canister can be provided with single or double lid design. Each lid is welded remotely following fuel loading and drying of the canister. In the particuler case of the flanged design, a single lid is bolted to the closure flange.

For the flanged design, two penetrations are provided through the lid, one for draining and the other for venting. A double-seal mechanical closure is provided for each penetration. The lid is fastened to the body by bolting. Double metallic O-ring seals with interspace leakage monitoring are provided with the lid. The monitoring system developed by TNI for the MACSTOR System is virtually identical to that used on its large metal casks. This system incorporates a simple, rugged, pressure maintenance system using helium within the metallic seal interspaces of all canister penetrations. Through the monitoring of these pressures and record keeping using operator logs or computerized printout, the utility is able to demonstrate that the spent fuel helium environment has been maintained over the period of storage. This provides substantial safety assurance and eliminates fuel inspection.

The Transfer Cask

The transfer cask design is based on conventional transport cask designs and European hot-cell transfer cask technology. It is adapted to the unique requirements for on-site movements consisting of rapid and simple loading and unloading.

The transfer cask is loaded with the sealed canister through the top and unloaded through the bottom. Please refer to Figure 3.

The MACSTOR Loading System

The MACSTOR Loading System consists of two major components: the gantry system and the loading platform. The gantry system is used to lift the transfer cask and the fuel canister. The loading platform is used as a mating shielding structure between the transfer cask bottom plug and the top of the concrete module.

The Gantry System consists of a rail-mounted bridge with double girders on top of the MACSTOR Module. It is equipped with a main trolley and an auxiliary trolley.

The main trolley carries the transfer cask hoist fitted with a lifting beam and a auxiliary lifting mechanism consisting of a canister hoist and its pneumatic grapple. The lifting beam and the pneumatic grapple are special lifting devices designed, manufactured and tested in accordance with the requirements of ANSI 14.6-1986.

The auxiliary trolley carries a 10-ton hoist which handles the cavity shield plug during fuel loading.

The Transporter Vehicle and Auxiliary Equipment

A specially designed heavy-haul trailer with trunnion support pedestals is used to transport the transfer cask from the fuel building to the module. Auxiliary equipment such as a transfer cask lifting beam, a canister grapple mechanism and a vacuum drying system are also supplied.

MASCTOR Fuel Handling Operations

The fuel handling operations can be divided in two main parts: the pool operations and the storage site operations. The pool operations are quite similar to most cask loading operations performed at pools and illustrated in Figure 4.

Exposure of operating personnel to radiation is kept to a minimum because:

- the transfer of fuel assemblies into a canister is performed underwater;

- all other operations (cask decontamination, canister draining and vacuum drying) are performed with the shielded canister closure lid in place.

The MACSTOR loading procedure is similar to the proven vertical loading operations currently in use. Please refer to Figure 5. Certain modifications have been adopted to simplify the handling of the cask bottom plug. The gantry system provides all the lifting capabilities required for loading the Module. The loading platform assures shielding during operations. Indirect radiation exposure (sky shine) is only present for a short duration during cask transfer. The fuel handling personnel momentarily evacuates the module top slab and can perform all crane operations from a distance with the crane pendant controls.

The storage capacity for one MACSTOR module is determined by the number of spent fuel assemblies held by each canister and by the number of canisters contained in each module. Please refer to Table 3.

Optimization of canister capacity required various factors such as fuel burnup, cooling time and weight to be taken into consideration. The selection of high burnup and less than 10 year cooling time required an efficient basket design to maintain allowable fuel temperatures during long term storage.

The inherent higher shielding requirements and good heat rejection capability increasing the weight of both the canister and transfer cask restrict the number of fuel assemblies that can be stored in each canister. The geometry of the canister and the transfer cask then evolved and defined a maximum lift requirement which cannot exceed 125 tons. Utilities can therefore implement the MACSTOR system without having to upgrade the pool crane lift capacity beyond 125 tons.

The module storage capacity, on the other hand, is an economic issue. The module can contain as few as eight canisters. Further canisters can be added in increments of four. This modularity allows adaptation to lower storage capacity with higher costs or increased storage capacity with lower costs.

There are, however, practical limitations to building large modules. The large concrete structure is more difficult to design and construct, and requires more reinforcing steel components.

Consequently, the optimum capacity for a module is considered to be 16 to 20 canisters.

4.   ADVANTAGES OF THE MACSTOR SYSTEMS

The MACSTOR System has been designed and tested so that a single system can be utilized for all fuel discharged at a given site. When storing high burnup fuel, concrete storage systems with less-than-optimal thermal design will have problems with concrete and fuel temperatures. The MACSTOR System was developed with high volumetric air flow rates which provide efficient cooling and lower fuel, canister, and concrete temperatures. MACSTOR technology is therefore particularly well suited for high burnup fuel and represents the design of the future for the storage of the next generation of spent fuel.

The MACSTOR System has a canister design that meets all the requirements of an ASME Code Section III, Class I vessel. The closure seal is based on proven metal cask seal technology and is designed with a monitoring system to assure the utility, the regulators, and the public, that plant operators can confidently establish and verify the safe and inert environment that is surrounding the fuel.

Because of the hydrid nature of the MACSTOR design, the manpower requirements and operator exposure occasioned by the loading and transfer of the canisters is very similar to that for metal casks. MACSTOR will require no more than half of the operations time and exposure on a pre-assembly basis as that for other modular designs.

One of the drawbacks of some concrete storage systems is that they require more land area than typical metal cask storage systems. Vertical storage was selected for the MACSTOR System, to maximize its space efficiency. Other "stand alone" concrete systems require more than 1.5-2.5 times the land area necessary for MACSTOR. Since most other costs associated with an Independent Spent Fuel Storage Installation (ISFSI) are proportional to land area, it is clear that the MACSTOR ISFSI is less expensive than those other "stand alone" concrete designs. MACSTOR is the most space efficient storage system available today.

The MACSTOR System has been developed for ease of transport licensing and for optimal interface with the eventual off-site transport of the spent fuel. With the MACSTOR system, no spent fuel pool is required to load a transport cask. The space between facing storage modules can be converted to a set down area where the transport cask is vertically positioned prior to loading. The transfer cask removes a canister from a storage location and transfers it into the transport cask using the MACSTOR loading system. The transport interface design of MACSTOR is second only to that of dual purpose casks.

5. SUMMARY

The most significant characteristics of the MACSTOR system are presented in the following summary.

The flexibility of the system facilitates application to store both LWR and CANDU spent fuel. The system can be equally used at a reactor site or at a centralized storage site due its interfacing capability with either the transfer or the transport cask. It is economical in terms of implementation, operational cost and storage space utilization.

The storage methodology and component design are based on proven features and experience obtained during the past twenty years. Operational feedback has been positive since the first implementation in 1995 at Gentilly 2 NGS.

The simplicity of the MACSTOR system allows utilities to execute all construction, purchasing and fuel transfer operations. The local economy in the region of the storage facility draws direct benefits through participation in the construction activities on a regular and continuous basis.

6. REFERENCES

- Pattantyus, P., "The Transnuclear/AECL MACSTOR System for Spent Fuel Storage" (INMM Spent Fuel Management Seminar IX, Washinton, D.C., 1992).

- Paré, F.E., Pattantyus, P., Hanson, A.S., "MACSTOR: Dry Spent Fuel Storage for the Nuclear Power Industry" (Proc. International, Conference on Nuclear Waste and Environmental Remediation, Prague, Czech Republic, 1993).

- Cloutier, L., Pattantyus, P., "Advances in Spent Fuel Storage Technology for CANDU Reactors" (IAEE Technical Committee Meetings on advances in Heavy Water Reactors, Toronto, Canada, 1993).

- Girard, A.-M., Cloutier, L., Macici, N., Moussalam, E., "Implementation of a Spent Fuel Dry Storage Facility at Gentilly 2" (Proc. CNA/CNS Annual Conference, Montreal, 1994).

- Paré, F.E., Joubert, W.-M., "Evolution of the MACSTOR Dry Spent Fuel Storage System" (Proc. International Symposium on Safety and Engineering Aspects of Spent Fuel Storage, Vienna, Austria, 1994).

- Moffet, R., Sabourin, G., Cherradi., "Prescon 2 Simulation of MACSTOR tests (Proc. CNA/CNS Annual Conference, Montreal, 1994).

TABLE 1 AECL EXPERIENCE IN DRY FUEL STORAGE PROJECTS

Site & Location	Reactor Size & Type	Quantity of Spent Fuel	Number of Canisters	Capacity of Canister	Status
Whiteshell Laboratory Pinawa, Manitoba Operationg since 1975	Prototype Organic-cooled 55 Mw	23 MgU	11	Variable	Complete
Gentilly-1 Station Gentilly (Quebec) Operating since 1985	Prototype BLW 250 Mw	67 MgU	11	304	Complete
Douglas Point Station, Kincardine, Ontario Operating since 1987	Prototype PHW 200 Mw	298 MgU	47	486	Complete
NPD Station Rolphton, Ontario Operationg since 1989	Prototype PHW 25 Mw	75 MgU	12	486	Complete
Lepreau 1 Station Point Lepreau, New Brunswick Operating sinice 1991	Commercial PHW 600 Mw	2800 MgU (over 30-year lifetime)	91	540	Operating
Wolsong 1 Station Korea Operating since 1992	Commercial PHW 600 Mw	2800 MgU (over 30-year lifetime)	91	540	Operating
Gentilly-2 Station Gentilly, Quebec Operating since 1995	Commercial PHW 600 Mw	2800 MgU	-	540	Operating

TABLE 2 MACSTOR SYSTEM CONTAINMENT TYPES

	Thermal / Capacity Performance		
	Low	Medium	High
Containment type	Double	Single	Single
Thermal rating	up to 6 kW	up to 9 kW	up to 24 kW
Heat shield around containment	not required	not required	required
Canister Closure System	Welded (single lid)	Bolted (single lid) Welded (double lid)	Welded (double lid)
Cover gas	Helium	Helium	Helium
Cover gas monitoring method	Indirect through secondary containment	Seal or lid interspace	Lid interspace

TABLE 3   TYPICAL DESIGN PARAMETERS OF AECL'S MACSTOR STORAGE SYSTEM

COMPONENT	PWR (reference)	BWR (reference)	VVER 440 (reference)	VVER 1000 (reference)
Fuel type	CE 16 x 16 System 80	General electric 4,5,6 assembly class	VVER 440	VVER 1000
Fuel length	4528 mm (178.3 in.)	4474 mm (176.16")	3217 mm (126.6 in.)	4584 mm (180.5in.)
Fuel active length	4094 mm (161 in.)	4160 mm (163.8 in.)	2420 mm (95.3 in.)	3630 mm (142.9 in.)
Fuel cooling period	> 7 years	> 7 years	> 6 years	> 10 years
Fuel enrichment	2.6% U 235	3.67 % U 235	3.6 % U 235	3.7 % U 235
Fuel element thermal rating	1 kW	400 W	300 W	1 kW
Canister capacity (fuel assemblies)	21	44	55	19
Canister thermal	21 kW	18 kW	16.5 kW	19 kW
Fuel burn-up (nominal)	40 GWd/MTU	40 GWd/MTU	40-47 GWd/MTU	49 GWd/MTU
Fuel Uranium mass	413 kgU	195 kgU	120 kgU	486 kgU
Transfer cask weight	110 tons	110 tons	80 tonnes	110 tonnes
Module Capacity	16 canisters	16 canisters	16 canisters	16 canisters
Height	6.6 m (21.6 ft)	6.6 m (21.6 ft)	5.5 m (18 ft)	7 m (23 ft)
Width	9 m	9 m	9 m	9 m
Length	22 m	22 m	22 m	22 m
MACSTOR module material	regular density (150 pcf) reinforced concrete	regular density (150 pcf) reinforced concrete	regular density (150 pcf) reinforced concrete	regular density (150 pcf) reinforced concrete
Canister material	Carbon steel	Carbon steel	Carbon steel	Carbon steel
Canister classification	ASME Section III Subsection NB	ASME Section III Subsection NB	ASME Section III Subsection NB	ASME Section III Subsection NB
Cover gas	Helium	Helium	Helium	Helium
Dose rate:				
Walls	< 25 micro Sv/h	< 25 micro Sv/h	< 25 micro Sv/h	< 25 micro Sv/h
Top slab	< 25 micro Sv/h	< 25 micro Sv/h	< 25 micro Sv/h	< 25 micro Sv/h
Air inlets and outlets	< 200 micro Sv/h	< 200 micro Sv/h	< 200 micro Sv/h	< 200 micro Sv/h

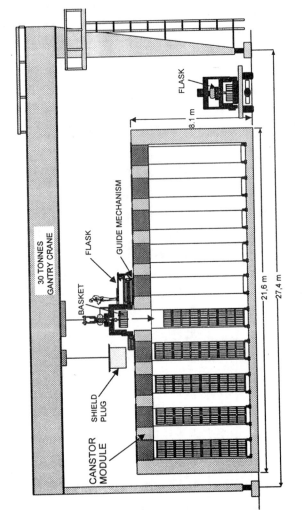

FIGURE 1   CANSTOR MODULE

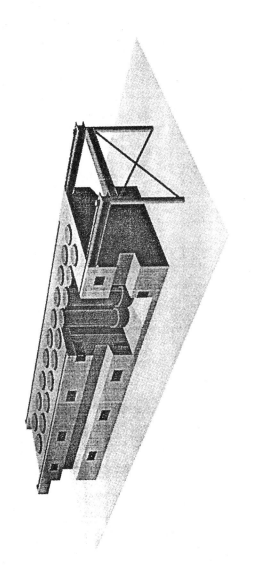

FIGURE 2   MACSTOR MODULE

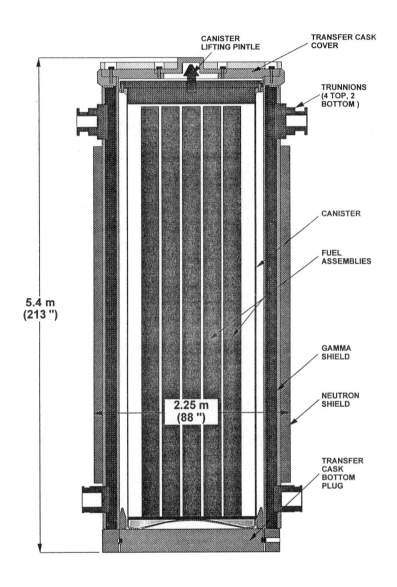

CANISTER
LIFTING PINTLE

TRANSFER CASK
COVER

TRUNNIONS
(4 TOP, 2
BOTTOM)

CANISTER

FUEL
ASSEMBLIES

GAMMA
SHIELD

NEUTRON
SHIELD

TRANSFER
CASK
BOTTOM
PLUG

5.4 m
(213 ")

2.25 m
(88 ")

FIGURE 3   TRANSFER CASK

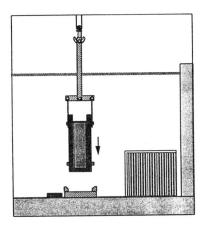

LOWER TRANSFER CASK
(with fuel canister inside)
IN SPENT FUEL POOL

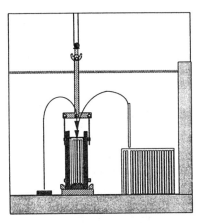

**TRANSFER FUEL ASSEMBLIES
IN TRANSFER CASK, POSITION
CANISTER LID AND CASK LID**

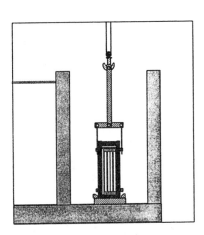

**MOVE TRANSFER CASK TO
DECONTAMINATION AREA, DRAIN,
DRY AND BACKFILL WITH HELIUM**

**LOAD TRANSFER CASK ON
CASK TRANSPORTER**

FIGURE 4   POOL OPERATIONS FOR MACSTOR SYSTEM

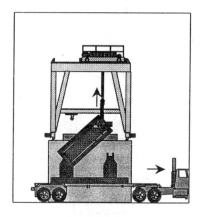

LIFT CASK ON TOP OF
MACSTOR MODULE

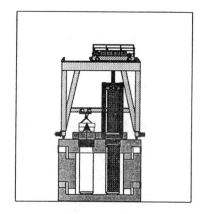

MOVE CASK ON
LOADING PLATFORM
UNLATCH BOTTOM PLUG

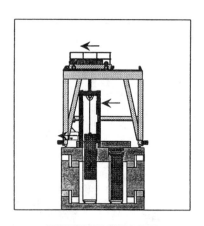

REMOVE CASK SHIELD PLUG,
MOVE TRANSFER CASK
AND LOWER CANISTER

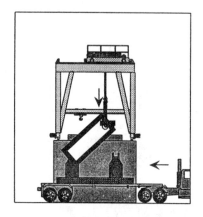

REPLACE CASK SHIELD PLUG
MOVE TRANSFER CASK
BACK TO TRANSPORTER

FIGURE 5   **STORAGE SITE OPERATIONS FOR MACSTOR SYSTEM**

pent nuclear fuel storage at a US utility

D DELGEORGE, D ELIAS, T J RAUSCH, A G PANAGOS, and K S PETERSEN
ComEd, Illinois, USA

I.   Introduction

Thank you for the opportunity to discuss this important matter with you today. At the outset, it is worth noting that the High Level Waste (HLW) issue has become critical in the United States of America not because of any action by U.S. utilities; but rather, because of the inaction of the U.S. federal government to meet an obligation it has acknowledged for over thirty years. It is often forgotten that the U.S. federal government made a commitment to disposition high level radioactive waste in order to promote the commercialization of nuclear energy. However, it seems evident that, the U.S. federal government has let its citizens people down, allowing a program for the orderly and cost effective disposal of nuclear waste to deteriorate into a bureaucratic nightmare and a swirling drain of consumer dollars.

ComEd has its headquarters in Chicago, Illinois. The company provides service to 8.2 million residents in the northern fifth of the State of Illinois. Annual revenues are $6.3 billion from the sale of 85 million MWH. The system is summer peaked, with demand on hot days near 19,000 MW. ComEd's total capacity is 22,523 MW divided almost equally between fossil and nuclear. We have 13 reactors at six nuclear sites. Five of the sites have two reactors while the sixth site, Dresden, has three. Twelve of the thirteen reactors are still operational and provide roughly 70% of ComEd's generation requirements. ComEd's first reactor, Dresden Unit 1, was the first commercial reactor built in the U.S. Dresden Unit 1 started operational testing in 1959 and obtained full power operation in 1960. Dresden Unit 1 was shutdown in 1979 and finally retired in 1983.

Based on the U.S. federal government's commitment to accept and disposition high level waste, ComEd and other U.S. utilities designed and built their plants expecting only limited temporary storage and on-going reprocessing of spent fuel. When it became clear during the Carter Administration (1976-1980) that U.S. federal government had changed the rules, ending our national commitment to reprocessing, these temporary storage facilities were enlarged. ComEd more than doubled the spent fuel pool storage capacity at our six nuclear plant sites, expecting the U.S. federal government to meet its restated obligation to begin acceptance of spent nuclear fuel for disposal by 1998, as was provided with the enactment of the 1982 Waste Policy Act. If the U.S. federal government meets this obligation to accept spent fuel in 1998, ComEd and our ratepayers would incur no additional costs for interim storage of spent fuel. If U.S. federal government acceptance of fuel is delayed until 2015, ComEd expects to spend over US$ 1.2 billion to further expance spent fuel storage.

As a result, Illinois consumers face the very real prospect of paying for additional on-site storage facilities and the legal costs of fighting a protracted firestorm of anti-nuclear and environmental protests surrounding the licensing of additional on-site storage capacity. For these reasons, ComEd is strongly supporting legislation to force the U.S. federal government to accept spent nuclear fuel by 1998 as mandated by the law.

II.   Dresden 1 Storage Considerations

Dresden Station consists of three units at a site located about 80 kilometers southwest of Chicago, Illinois. Dresden Unit 1 is a 700 MW-thermal (207 Mwe) boiling water reactor (BWR) that operated from 1960 to 1978. It is presently retired from service. Dresden Units 2 and 3 are 2527 MW-thermal BWRs (834 Mwe) which have been operating for about 25 years. Dresden Unit 1 was shutdown for chemical cleaning and modifications in 1978. Post Three Mile Island modifications were found to be not cost effective, and it was decided to decommission the unit in 1983. Although several portions of the plant containing radiological source terms are being dismantled and disposed of, most of the plant is being maintained in a safe storage (SAFSTOR) condition.

Dresden Unit 1 has a spent fuel storage/transfer pool that is separate from the larger units. It is constructed on bedrock and made of several foot-thick concrete that has a painted-on epoxy liner. It presently contains 683 spent nuclear fuel assemblies. The fuel is slightly smaller than a standard BWR assembly. It has a nominal 4.85 inch pitch versus 6.0 inches for typical BWRs, and has an active fuel length of 110 inches versus the typical 145 inches.

In the 1960's, ComEd shipped 897 assemblies to a fuel reprocessing plant located in West Valley, NY. Most of the spent nuclear fuel was reprocessed. The resulting plutonium and enriched uranium was sold back to the U.S. federal government. Prior to reprocessing 206 assemblies were returned to the Dresden site, following the U.S. federal government ban on reprocessing. These spent nuclear fuel assemblies are currently stored in the Dresden Unit 2 and 3 spent fuel pools.

In late 1993, ComEd concluded that it would be prudent to remove the fuel from the Dresden Unit 1 spent fuel pool for two major reasons. First the risk of leakage is increasing with pool age. Second, the fuel racks consist of overhead supports bolted to the pool walls, and long term rack bolting corrosion levels are a concern. To ensure continued safe operation of the pool until it can be emptied, corrosion samples are under surveillance, and the groundwater in the area surrounding the pool is being monitored for radioisotopes.

ComEd considered several available alternatives for emptying the Dresden Unit 1 spent nuclear fuel pool, including pool refurbishment (e.g. stainless steel lining of the pool), relocation to another ComEd reactor pool, dry storage, or relying on the U.S. federal government. We concluded that dry storage is the preferred approach. The cost of pool refurbishment far exceeded the estimated costs of dry storage, and would still require significant additional handling of the more than 20 year old spent nuclear fuel. Options ComEd has available for relocating the fuel is to either move it to Dresden Units 2 or 3 pools, or ship it to one of our other 5 nuclear stations. The relocation option was considered a viable contingency, should problems in the Dresden Unit 1 pool develop, but much less desirable than dry storage. Although all of ComEd's operating reactor spent fuel pools have several or more years space available, none have sufficient space to operate to license expiration. Relocation would, therefore, only hasten the need for dry storage or fuel pool expansion at one of the other operating reactors, while incurring the additional expense of handling and transporting the spent fuel.

Dry storage has already been proven economical in several domestic applications, and a general licensing approach was available to minimize the potential for installation delays. A dry storage system could be designed consistent with the Multi-Purpose Canister (MPC) system being developed by the U.S. federal government at that time. This would maximize the probability that the fuel could be moved to a U.S. federal repository without being re-canistered. In addition, we could procure a system independently of the U.S. federal government to ensure that the spent nuclear fuel is removed from our spent fuel pools expeditiously. Once the spent fuel pool is emptied, ComEd anticipates substantial cost savings in operations and maintenance.

III.   Dry Storage Selection for Dresden Unit 1

ComEd evaluated a number of vendors qualified to provide spent nuclear fuel dry storage systems.

The U.S. federal governments' Multi-Purpose Canister (MPC) specification was used to establish many of the requirements. This approach would maximize the probability that the fuel could be stored on-site and subsequently transported to an off-site storage area and/or eventually to a U.S. federal repository. By chosing a system that is fully compatible to the MPC specification the chance of a dry cask storage system not being accepted by the U.S. federal government would be minimized.

Schedule and cost considerations also played an important part in the evaluation. The system selected needs to be licensed by the U.S. Nuclear Regulatory Commission on a schedule compatible with ComEd's plan to remove the fuel from the Dresden Unit 1. Handling and repackaging of spent nuclear fuel also had to be minimized. The design needed to be flexible and generic consistent with the need to execute similar projects in the future at other nuclear sites. The overall cost of storing spent nuclear fuel needed to be minimized. At the time of the evaluation the U.S. federal government had not yet selected the vendor for their MPC system.

During the bid evaluation phase, each bidder was graded for capability in the following areas:

EVALUATION AREAS

- MPC Compatibility
- Technical Competency
- Manufacturing Capability
- Operations, Maintenance & Decommissioning
- Project Management
- Licensibility Of Design
- Life Cycle Costs of the Storage System

Our evaluation found Holtec International and their fabricator, U.S. Tool and Die (UST&D), to be the most acceptable bidder due to the following system attributes:

EVALUATION RESULTS

- Highest Capacity MPC Design (68 Assemblies)
- Fully Committed To MPC Approach
- MPC Design Submitted To U.S. NRC for Review
- Licensing Schedule at Least one year ahead of U.S Federal Government MPC Schedule
- Holtec/UST&D is a Strong Team
- High Marks for Technical Capability and Manufacturing Respectively
- Proven Track Record in the Design & Manufacturing of Spent Nuclear Fuel Pool

IV.   System Design Summary

The selected system is named the HI-STAR (Holtec International Storage, Transportation and Repository) system. The HI-STAR Topical Safety Analysis Report is under review by the U.S. NRC, docket number 72-1008, for certification under 10CFR72, Subpart L. In addition, a Safety Analysis Report for Packaging (SARP) has been submitted to the U.S. NRC for a transportation Certificate of Compliance, docket number 71-9261. Holtec has developed the HI-STAR System to be fully compliant with the U.S. federal government requirements on MPC systems.

It was the only dual-purpose MPC system under full NRC review at the time of the evaluation with the capability for both storage and transport using a single overpack. Additionally, the system includes a unique option: a concrete storage-only module for those utilities that wish to purchase a storage only system. This feature, termed the Holtec International Storage and Transfer Operation Reinforced Module (HI-STORM), serves to replace the multi-purpose metal overpack of the HI-STAR System.

HI-STAR is currently being licensed to either store or transport 32 PWR spent fuel assemblies with burnup credit, 24 PWR assemblies without burnup credit, and 68 BWR assemblies with or without burnup credit. Each MPC design can be safely stored with fuel cooled for five years and can be transported with fuel cooled for ten years.

The HI-STAR System can be used for on-site spent nuclear fuel storage and/or off-site shipment of spent fuel. If the HI-STORM system is used, a transfer cask system can exchange the MPC between storage and transport casks. Figure 1 shows the HI-STORM with the MPC.

The HI-STAR MPC is an assembly consisting of a honeycomb fuel basket, bottom plate and canister shell. By building the basket from an array of edge-welded square boxes, a honeycomb structure is created which has the inherent strength to susten drop accident loads in excess of 60g laterally and more than 100g vertically. The construction of the MPC mirrors the process utilized by Holtec in the fabrication storage racks for in-pool storage of spent nuclear fuel. Figure 2 is representative of the several MPC designs.

The HI-STAR System metal cask is an extremely rugged, heavy-wall, cylindrical vessel with an integrally welded bottom and a bolted-top closure lid. The chief constituents of the cask are a thick inner shell welded to a thick bottom plate at the base and a heavy forging at the top. The bolted lid configuration provides protection for the closure bolts and gaskets in the event the cask were to experience a severe impact. Additional layers of carbon-steel plate are placed around the inner shell to form a protective barrier and serve as an effective gamma radiation shield. The outer shell of the storage cask provides excellent neutron shielding.

The cask utilizes layered pressure vessels. The layered construction also provides added strength from improved ductility and elimination of the potential for thru-wall cracks due to material flaws or impacts. The HI-STAR System dual purpose cask offers transport-compatible materials for the confinement boundary, outer shell, and bolts. Figure 3 is representative of the metal casks in the HI-STAR System.

V.   Current Status/Plans and Conclusions

Dresden Unit 1 is a shutdown reactor and therefore, we do not have the pressure of maintaining adequate spent nuclear fuel pool space for continued operation. Under such pressure, most United States utilities have chosen systems already licensed for storage, but that have not been licensed for transportation.

ComEd has taken the long term view by procuring a system that is not only dual-purpose (storage and transportation), but is also being designed to the U.S. federal government MPC specification. We believe that it is possible to achieve this at a reasonable cost. That upon ultimate shipment to the U.S. federal repository, this will be the lowest cost approach, because the fuel will not have to be repackaged for transportation. It is too early to predict precisely what the canister requirements will be for the repository. By designing and building it to the MPC specification, resulting in the use of high quality stainless steels and long service life design features. Thus the probability of the MPC and its contents ultimately being acceptable in the repository is increased.

Our actions on Dresden Unit 1 will result in the licensed availability of dual-purpose systems for our operating units, some of which begin to lose full core discharge capability in the early 2000's. The HISTAR / HISTORM systems are designed to handle all of the fuel currently in ComEd's spent nuclear fuel pools. The experiences we are gaining in implementing dry storage for Dresden Unit 1 will prove invaluable should we decide to pursue dry storage for these units. The procurement of transportation HI-STAR overpacks for Dresden Unit 1 fuel provides ComEd with a transportation readiness.

The experiences of other U.S. utilities, and extensive discussions with our regulators have taught us the importance of planning far ahead, and the need to continually learn from others. These utilities have had to face major hurdles such as formal Public Utility Commission approval, state legislative authorization, and formal public opposition/intervention in the licensing process. They and others have also been challenged by a number of generic Quality Assurance, licensing, and engineering issues which are now being resolved under an a formal U.S. NRC Action Plan. As a result ComEd is making special efforts in the areas of public communications, QA oversight of design and manufacturing, and the quality of site-specific design, modification and licensing documentation.

In conclusion, we have chosen to install a flexible and robust dry storage system at Dresden Unit 1. We are confident that the extra efforts associated with licensing, and ensuring a flexible design and a high quality manufactured product will pay dividends in the near term for Dresden Unit 1. We expect it will also pay dividends for our other operating units.

C512/0

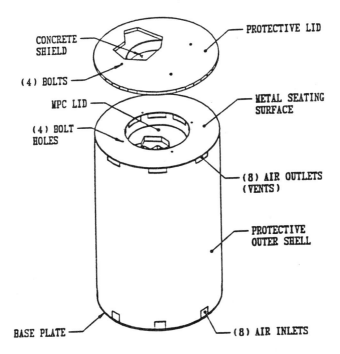

CONCRETE SHIELD

PROTECTIVE LID

(4) BOLTS

MPC LID

METAL SEATING SURFACE

(4) BOLT HOLES

(8) AIR OUTLETS (VENTS)

PROTECTIVE OUTER SHELL

BASE PLATE

(8) AIR INLETS

Figure 1: HI-STORM System

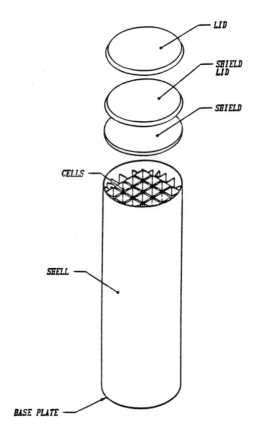

Figure 2: HI-STAR Canister

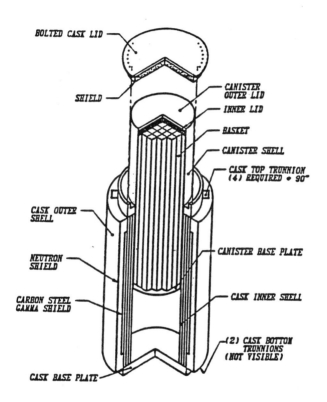

BOLTED CASK LID

SHIELD

CANISTER OUTER LID

INNER LID

BASKET

CANISTER SHELL

CASK TOP TRUNNION (4) REQUIRED ● 90°

CASK OUTER SHELL

NEUTRON SHIELD

CARBON STEEL GAMMA SHIELD

CANISTER BASE PLATE

CASK INNER SHELL

(2) CASK BOTTOM TRUNNIONS (NOT VISIBLE)

CASK BASE PLATE

Figure 3: HI-STAR System

Ensuring safe storage of spent nuclear fuel from NPPs in Russia

V KOZLOV, T F MAKARCHUK, V V MOROZOV, V V SPICHEV, and N S TIKHONOV
Russian Project and Scientific Research Institute of Complex Power Technology, St Petersburg, Russian Federation

Synopsis – Spent fuel (SW) management in Russia is considered. Basic safety ensurance principles of SF storage are presented. Status of SF storage in the country is shown. Examples of spent fuel storage facilities are given.

Nine nuclear power plants with 29 power units using WWER-440, WWER-1000, RBMK-1000, BN-600 and EGP reactors are now in operation in the Russian Federation (I).

Annual spent fuel arising is approximately 800 t U.

Russia now pursues the closed fuel cycle concept. This includes SF reprocessing, U and partially Pu recycling in thermal and breeder reactors, and vitrified radwaste storage in above ground storage facilities at reprocessing plants.

The spent fuel from WWER-440 and BN-600 is reprocessed at the reprocessing plant RT-I in the Urals, whereas the spent fuel from WWER-1000 reactors will be reprocessed at the plant RT-2 which is under construction near Krasnoyarsk.

The management of RBMK spent fuel is exceptional in that this fuel type is not meant for reprocessing because of its low fissile content.

However cooling at reactor pools and intermediate storage at the NPP and reprocessing plant sites or in a specialized storage facility are necessary steps in all the modes of spent fuel management.

The basic safety ensurance principles and status of SF storage in the country are considered in the paper.

It is common practice in Russia to use water pools for storing of spent NPP fuel.

Wet storage proved itself well in the initial period of nuclear power development and will remain dominate in the next decades.

The following technical requirements are imposed upon SF storage by regulatory documents(2, 3, 4, 5, 6) now in force in the Russian Federation.

- placing fuel in a pool while providing nuclear safety in storage and handling operations;
- radiation safety of the personnel during the fuel storage period according to the existing standards;
- fuel conditioning monitoring, organization of security in the storage facility;
- pool water cooling to the temperatures no higher than 50°C with decay heat removal;
- cleaning pool water by removing radioactive and other contaminants to provide the required water clarity for remote fuel handling under water;
- the storage structure eliminating water leaking into the environment, and localization of leaks;

- prevention of radioactivity release through the ventilation system of the storage facility by using purification filters;
- proper planning of transport operations for fuel dispatch from the storage facility;
- possibility to store cassettes with defective fuel assemblies (FA).

To meet these requirements, the following principle engineering decisions were used at the design and construction of AR at Reactor, AFR (RS) away from reactor (reactor site) and AFR (OS) away from reactor (off site) storage facilities:

- adequate FA arrangement in the pool, a water layer above the fuel and the pool wall thickness satisfying the nuclear and radiation safety requirements;
- pool equipping with systems for water cooling and purification, required water level, maintenance above water space ventilation, process and radiation control;
- stainless steel lining on the inner pool surfaces;
- collection of leaks from under the pool lining and their control to prevent ingress into the neighbouring rooms or ground;
- storage of FAs with defective fuel rods in sealed cans.

Radiation safety and environmental protection are also achieved by the following measures:

- location of the building in the radiological controlled area;
- zoning of rooms and the provision of a change room and air lock or barrier;
- liquid and solid radwaste collection and removal;
- decontamination system for transport and handling equipment and vehicles;
- radiation monitoring both inside and outside the building.

Possible emergency situations within the building are taken into consideration in the storage design, such as:

- drop of a loaded container;
- drop of a loaded basket;
- pipe rupture of the pool cooling system;
- loss of pool water;
- power supply loss.

In connection with the work aimed at the increase of the overall safety of nuclear power facilities, the consequences of beyond design basis accident are considered for the systems if SF storage such as SSCR (self-sustaining chain reaction), full loss of pool water, drop of the process equipment and structures on the pool bay floor or onto the fuel stored.

After the discharge from the reactor, spent fuel is stored at AR pools for 3-5 years. In some cases AFR/RS facilities are built for accumulation and storage of the spent fuel during a longer period more than 10 years.

AFR/RS facilities are built at all nuclear power plants with RBMK reactors and at the Novovoronezh NPP.

All storages are reinforced concrete stainless steel lined structures. Fuel is stored either at the bottom of the pool (WWER, BN) in baskets and in racks or hung from a metal roof structure (RBMK). The facilities are equipped with a full set of mechanisms necessary to receive, store and retrieve spent fuel. Types of wet storage facilities are shown in Fig.1, 2, 3.

Operation experience with water cooled storage pools shows that non-defective fuel with average burnup has maintained its cladding integrity and high corrosion resistance for a long period of time. After several years of storage defective fuel elements have also shown no significant degradation.

Investigation of structural material behaviour in pool water (7,8) showed no significant change in the mechanical properties of zirconium cladding.

Stainless steels used for pool lining may also be classified as completely corrosion-resistant materials.

The data on storage facility quantities with the spent fuel from the major reactor types as of st August 1995 (I) are listed in Table I.

As can be seen RBMK spent fuel quantities present considerable difficulties. Dispatch of he spent fuel has not been carried out.

The storage capacity of operating RBMK facilities (including that obtained lay denser FA arrangement will provide SF reception from NPPs up to 2004. The above situation points to he pressing problem of long-term storage.

Dry storage alternatives have also been proposed in view of the necessity for long-term storage of RBMK-1000 spent fuel.

They have some merits in comparison with wet technology.

A method has been developed for storing RBMK fuel in SS canisters placed into massive concrete structures or vaults. In view of the need for long-term storage such facilities shall provide for:

- spent fuel security during a period no less than 100 years;
- proper temperature regime (below 200°C) that does not require maintaining an inert storage environment;
- retrievability of canned fuel from the facility for investigation purposes, including those connected with IAEA guarantees;
- precluding the access of atmospheric air to the constructional materials of fuel assemblies and cans;
- durability of the facility construction fro a period no less than 100 years;
- passive residual heat removal;
- capability for decontaminating hot surfaces in an operating storage facility;
- stability to external impacts (airplane crash, air blast, earthquake, flying objects, hurricane, tornado);
- quick and conventional identification of radioactive contamination source;
- minimum constructional expenses.

Fuel is stored in a massive ferroconcrete vault structure with special pits of 700mm diameter. The pits accommodate 2 or 3 tiers of sealed cans (diam. 630mm, height 4000mm) with 31 fuel assembles per can. Between the pits there are draught channels for cooling air circulation and heat removal from the ferroconcrete structure. Cooling air intake is achieved through the side walls of the storage building. The air flow is then directed into the compartment under the concrete structure which functions as an air distributing collector. Heated (in the draught channels) air goes up to the facility hall and further vented out through the roof into the environment.

The layout of a regional storage facility for RBMK-1000 spent fuel is shown in Fig. 4.

The storage area (designed for receiving spent fuel from all Russian and Ukrainian NPP with RBMK reactors) consists of 14 similar reinforced concrete vaults of sizes 18000 x 24000 x 14000 mm which contain 468 vertical shafts with blind ends. Vaults of the same type can be constructed at the NPP site.

An alternative to concrete vault structures are concrete casks.

A dual purpose cask (Fig. 5) is designed for long-term RBMK spent fuel storage at the NPP site.

The cask should meet the following requirements:

- compliance with the current national and IAEA rules and regulations regarding spent fuel transport and storage;

- service life of loaded containers no less than 50 years . Possible future usage will depend on the container and fuel condition after certain storage periods which will be defined later,
- double sealing barrier;
- gamma- and neutron radiation protection;
- possible use of current baskets and higher capacity baskets;
- air or inert gas as the filling medium of the cask inner cavity;
- maximum cladding temperature under normal storage conditions is no more than 200°C; periods at 380°C should not exceed 24 hours;
- fuel assemblies are stored disassembled in halves;
- reasonable costs;

The cask meets all the requirements of the current engineering standards for normal and emergency conditions. The shielding and strength properties of the cask are ensured by using concrete of high strength (strength > 100MPa and density > 4t/cubic metre) together with three sealing shells and reinforcing framework.

The cask has two lids with shock-resistant and metal-graphite gaskets.

The cask is transported in a shock-absorbing jacket. Under accidental conditions this damper relieves stress on the cask and provides additional shielding.

Characteristics of the cask are as follows:

Cask capacity	6.5 tU (57 fuel assemblies disassembled into 114 halves)
SF cooling period	5 years
Heat release (maximum)	9.4 kW
Maximum temperature on cask surface	< 85°C
Maximum dose on cask surface	< 60 mRem/hr (600 mSv/hr)
Cask mass:	
empty cask	70 t
loaded cask	85 t
loaded cask in transportation jacket	110 t
Cask dimensions:	
outer diameter	2300 mm
height	5125 mm

While developing the dry storage options the safety analysis was performed by the General Designer of the facility. The safety analysis incorporates preparing a list of initiating events for design and beyond-design basis accidents, justifying storage system safety for normal operation and design-basis accidents, assessing the course of beyond design basis accidents and developing a complex of compensatory measures.

The approval of the 'Safety analysis report' by the State Committee on Supervision for Nuclear and Radiation Safety of the Russian Federation (GOSATOMNADZOR) means confirmation of the technical decisions taken.

References

1. Kurnosov V.A., Makarchuk T.F., Morozov V.V. ,Tikhonov N.S. Spent Fuel Management in Russian Federation: State of Art and Outlook for Future. Meeting of Regulary Advisory Group on Spent Fuel Management. IAEA, Vienna, Austria, September 25-29, 1995.

2. General Rules of NPP Safety Provision. OPB-88. PNAE G-OII-89.

3. General Rules and Regulations for Spent Fuel Storage and Transportation at Nuclear installations. PNAE G-14-029-01.

4. Sanitary Rules for NPP Design and Operation. SPA C-88.

5. Basic Sanitary Rules of Working with radioactive Substances and Other Sources of Ionizing Radiation. OSP-72/87.

6. Radiation Safety Standards. NRB-76/87.

7. Kondratyev A.N. , Makarchuk T.F., Morozov V.V. Tikhonov N.S. Status of the USSR Spent Nuclear Fuel Storage. International Seminar on Spent Fuel Storage – Safety, Engineering and Environment Aspects, IAEA, Vienna, Austria, 8-12.10.1990.

8. Ivashov Yu.V., Kolobov E.A., Makarchuk T.F., Novikov Y.B., Tikhonov N.S. Corrosion Testing of Construction Materials of Spent WWER-440 Fuel. Safety and Engineering Aspects of Spent Fuel Storage Proceedings of a Symposium, Vienna, 10-14.10.1994, IAEA-NEA (OECD), 193-199.

Table I. Filling of storage facilities with spent fuel from major types of power reactor

Type of reactor	WWER-440		WWER-1000				RBMK-1000			
Storage facility type	At-reactor		At-reactor		Away-from-reactor		At-reactor		Away-from-reactor	
	Storage capacity	Filling	Storage capacity	Filling	Storage capacity	Filling	Storage capacity	Filling	Storage capacity	Filling
Fuel assemblies, items	3900	2300	27600	II53	900	26	30I00 *	17850	35000	39580 *
Fuel quantities, tU	470	230	I200	505	400	II	3430	2040	3990	4500

* Storage capacity with dense storage mode.

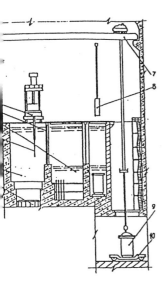

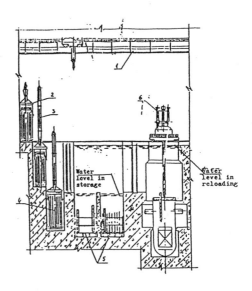

Fig.I. AR cooling pool for WWER-440 spent fuel

1-reactor; 2-reactor well; 3-FA; 4-cooling pool;
5-lock; 6-reloading machine; 7-crane; 8-basket;
9-cask for SFA; I0-truck.

Fig.2. AR cooling pool for WWER-I000 spent fuel

I-circular electric crane, 320+I60/2+70 lifting
capacity; 2-traverse for spent fuel cask; 3-rod
for cask; 4-shipping cask; 5-cooling pool racks;
6-reloading machine.

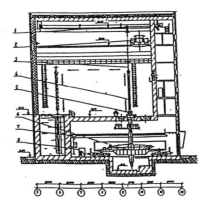

Fig.3. Interim storage for RBMK reactor spent fuel

    I - Cable trolley, I5t lifting capacity;
    2 - Overhead crane, 20/5t lifting capacity;
    4 - Transport transfer basket, 9 seats;
    5 - Guiding device;
    6 - Cask-car (modified TK-8);
    7 - Overhead track hoist;
    8 - Longitudinal alignment device.

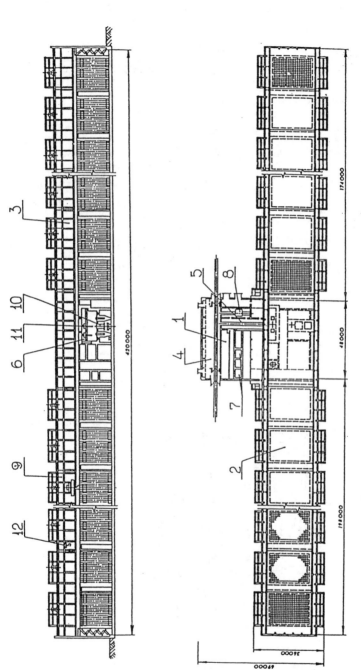

FIG. 4. REGIONAL STORAGE FACILITY FOR
RBMK REACTOR SPENT FUEL

I - cask reception; 2 - storage area; 3 - transport hall;
4 - cask-car reception; 5 - cask unloading; 6 - canning area;
7 - washing area; 8 - cask dispatch; 9 - loading/unloading
machine; IO - cask; II - manipulator; I2 - overhead crane

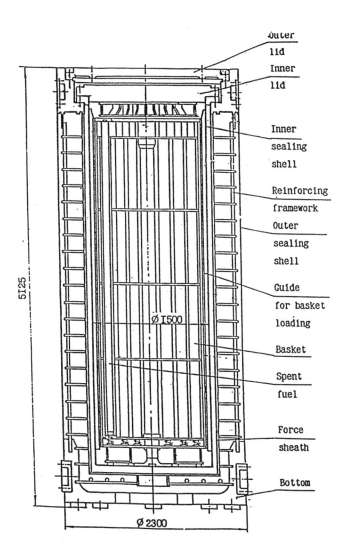

Outer
lid

Inner
lid

Inner
sealing
shell

Reinforcing
framework

Outer
sealing
shell

Guide
for basket
loading

Basket

Spent
fuel

Force
sheath

Bottom

Ø I500

Ø 2300

5I25

Fig-5 .METAL-CONCRETE CASK FOR SPENT FUEL OF RBMK-1000 REACTOR

ᵉ spent fuel storage facility at Dukovany and its initial eration

)TZEM and U KUTSCHER
:M GmbH, Alzenau, Germany
CK
ᵉar Power Plant Dukovany, Czech Republic
USCHKE
ᵉllschaft für Nuklear-Behälter GmbH, Essen, Germany

SYNOPSIS

In December 1995 the dry interim spent fuel storage facility at the Dukovany Nuclear Power Plant went into operation. After an international tendering process and comparison of the bids, the owner of the NPP, Czech Power Company (CEZ, a.s.), decided on the technical concept of dry cask storage offered by the German consortium BL-DUKOVANY GNS-NUKEM. After receiving the construction permit in summer 1994 and completion of the building phase in 1995, "hot loading" and start of operation was possible in December 1995. The loading process and the operation of storage has been performed without any problems. Within the next months approximately 10 casks containing 840 fuel elements will be stored.

INTRODUCTION

As in 1989 the nuclear fuel services from the former Soviet Union changed, the necessity of interim spent fuel storage became urgent in the Czech Republic. The owner of the Dukovany NPP CEZ, a.s. had to ensure additional spent fuel storage capacity starting in 1995, beyond the existing wet storage pools at the reactor. The tendering process started in February 1991 and was finished in 1992. Preliminary bids were submitted by 10 bidders from 7 countries. The proposed types of storage systems covered nearly all spent fuel storage methods. The technical concept of a dry cask storage offered by the consortium GNS-NUKEM fulfils all requirements and is very similar to the existing reference plants Ahaus and Gorleben in Germany. Therefore a short licensing and realisation time was expected. For the fuel type VVER 440 a new cask, the CASTOR 440/84, based on the design and safety principles of the CASTOR family had to be developed. With the preparation of the Site Permit Documentation including a Preliminary Safety Case, the race against time started.

DRY CASK STORAGE CONCEPT

The principle of the dry cask concept is based on the containment of the fuel assemblies in casks suitable for both transport and storage. The main characteristics of the CASTOR cask design are:

- The Cask body is made of ductile cast iron (DCI) to guarantee gamma shielding, leak tightness and protection against mechanical and thermal loads under both normal operational and hypothetical accident and test conditions.
- Neutron shielding is provided by polyethylene rods integrated in the cask body wall.
- External fins on the cask improved heat removal (22 KW).
- Double barrier system realised by the lid system consisting of primary and secondary lids sealed with metallic gaskets.
- Permanent leak tightness control is realised by a special pressure monitoring system.
- Qualification as a type B(U) package according IAEA regulations.
- The cask is independent from energy supply, ducted cooling or other systems.
- The suitability at the cask was demonstrated in a huge number of tests. The cask withstands the following accident conditions among others:
    * aircraft crash
    * drop of heavy loads
    * gas cloud explosion
    * fire
    * temporary burial
    * earthquake
    * crashes with other transport vehicles
    * drop

The building in which the casks are stored is all weather, ensures the handling capabilities and minimizes the impact of the radioactivity on the environment. It is built in a purely conventional way inside the area of Dukovany Nuclear Power Plant. The features of the storage building itself are as follows:

- Length 55 m
- Width 26 m
- Height 20 m
- Capacity 60 casks with a total of 5040 fuel elements
- Overhead crane with 130 tons lifting capacity
- Central control room for cask tightness surveillance and radiation control
- Cask maintenance room
- Road and rail connection
- Safety related data are transferred directly to the NPP control room

The CASTOR 440/84 CASK

The CASTOR 440/84 has been specially designed for the transport and the longterm interim storage of fuel assemblies from reactors of the type VVER 440 in operation in Eastern Europe. Cask dry storage using the CASTOR 440/84 is currently being realized at the Nuclear Power Plant Dukovany and Greifswald in eastern part of Germany.

The CASTOR 440/84 ca accomodate 84 fuel assemblies with an average initial enrichment of 3.5 % U235, an average burn-up of 33000 MWd/tHM and a cooling time of 5 years.

The hexagonal arrangement of this type of fuel assemblies and their large number required a special constriction fpr the fuel basket. The chosen design guarantees heat ellmination and criticallty safety under normal and testing conditions. The hexagonal tubes made of boratedsttel in the fuel basket have a honeycombed arrangement. Between the fuel positions of this arrangment, aluminium plates are adapted to the contours. In the radial direction, the plates have different lengths and assure sufficient heat elimination. The residual free spces are filled with small aluminium plates to guarantee sufficient strength.

LICENSING STAGES

The licensing stages for the facility and the cask type CASTOR 440/84 were as follows :

Site Permit	April 1993
Construction Permit	June 1994
Acceptance of facility	June 1995
Type B(U) approval of cask in Germany	June 1995
Validation of Type B(U) approval in CR	Nov. 1995
Preliminary Storage license (one year test period)	Nov. 1995
Operation License after one year test period	Nov. 1996 (expected)

The close cooperation between Czech and German Authorities, along with the very safe concept, provided the best conditions for a smooth licensing procedure of the plant in the Czech Republic. The erection period was limited due to the small storage capacities remaining in the wet pools. For the storage building, erected by domestic subsuppliers of the consortium, a completion time of about one year was allowed.

LOADING AND STORAGE OF THE FIRST CASK

Until November 1995, the old C30 cask type holding 30 irradiated fuel assemblies was the only cask type which could be used by the NPP Dukovany. The new cask CASTOR 440/84 providing higher capacity for 84 fuel assemblies of VVER 440 type and the appropriate handling equipment had to be introduced into the spent fuel system managment.

One month prior to the start of loading activities, cold handling trials were performedd with the new cask design.

Staffs of both the reactor and the supplier of the loading machine were involved to check the loading of the basket by means of a dummy. Additionally the exact coordinates of the cask in the pool were transfered to the software of the loading machine.

In accordance with a sequence plan, the reactor staff was trained in all handling steps which are necessary to prepare the cask for dry storage of irradiated fuel, such as the following:

- assembly of lid seals
- checking the cleanliness of sealing surfaces
- connecting and disconnecting of the lifting yoke arms at the pond
- draining of water inside the cask and sealing gaps
- screwing the lid bolts
- vacuum drying of the cavity
- performing the helium leak test on all gaskets
- back filling of the cavity with helium
- assembly of the pressure monitoring device
- transfer of the cask to the dry storage facility

After successfully testing, the Czech licensing authority SONS gave their permission for the hot loading of the cask, which began on November 15, 1995.

In the morning of November 16, the cask was already loaded with 84 fuel assemblies having a total of approx. 11 KW heat capacity.

Dose rate measurements were taken in the area of the upper trunnions. middle of sidewall and botton. The results are as follows:

Doses Rates Measured on the CASTOR 440/84 in the Reactor Building After Loading

1. 10 cm Distance from cask

Area	Gammas µSV/h	Neutrons µSV/h	Total µSV/h
Upper trunnions	50	50	100
Middle	45	50	95
Botton	13	200	213

2. 1 m Distance from cask

Area	Gammas µSV/h	Neutrons µSV/h	Total µSV/h
Upper trunnions	18	40	58
Middle	18	80	98
Botton	8	80	88

After draining, drying and helium filling of the cavity, the cask was stored in the reactor building waiting for temperature equilibrium. Thermocouples were fixed to different parts of the cask in order to provide temperature recordings. The maximum temperature was measured at the finned surface on the cask sidewall, amounting to approx. 57 °C.

Finally on December 5, a release was given to transfer the cask from the reactor building to the storage facility, which was done on the same day.

In the storage facility there were rather cold temperatures (under 0 °C) at the time of emplacement on the cask. Thus, the interlid pressure went down due to the temperature change from the reactor building. This is not a problem, but a law of nature.

Until now, no problems have been encountered in the storage, and certainly none are expected in the future.

idging the gap – the interim storage solution

ROBERTS BSc and **R A WARREN** BSc, MICE
FL Interim Storage Limited, Cheshire, UK

1   Synopsis

In Spent Fuel Management there are ultimately two available options, reprocessing/recycle and direct disposal.

Between reactor discharge and final destination lies a balance of economics and political considerations to be weighed carefully according to individuals needs. Common, is the necessity to allow heat producing wastes to decay before either option can be followed.

This is the principle behind 'Bridging the Gap', or more commonly, Interim Storage.

Interim Storage can be accomplished in a number of ways, dry in casks, in specially constructed vaults or within spent fuel pools, the exact method dependent upon utility or country policy.

Pool storage is analogous to fuel already held at Sellafield prior to reprocessing.

Interim Storage is not therefore a departure from a reprocessing policy since it is already an integral part of it, rather it is maintaining a deferred reprocessing option allowing for change in long term energy policy and Spent Fuel Management.

To expand the Company's portfolio and to enhance its range of back-end services, BNFL is assessing a number of technologies, from large vaults to metal dual-purpose cask and concrete cask storage systems. All these interim storage solutions have the potential for a direct interface with reprocessing plants should the customer select this fuel management strategy at a later date.

2   Background

Across the globe nuclear utilities face the same problem, what to do with spent fuel once it has been discharged from the reactor. There are only two long term options, namely reprocessing and direct disposal. With no proven direct disposal technology reprocessing has monopolised back-end fuel cycle strategies, particularly in countries with large nuclear programmes such as Japan, France, Germany and the UK.

For other countries, delaying a decision to commit to either reprocessing or direct disposal may seem to be the best option, at least until sufficient work has been done on the latter to have confidence in its costs and technical feasibility as a permanent solution. Delaying such a decision, resulting in a policy of interim storage, brings about a new set of problems, most notably what fuel storage strategy to adopt.

Essentially there are three options available to utilities: dry storage in purpose built casks, vault storage within specially constructed shafts or continued storage in a spent fuel pool.

This latter method is analogous to the manner in which fuel is already stored at BNFL's Sellafield site prior to reprocessing.

Interim storage should not therefore be viewed as a departure from a reprocessing policy since it is already an integral part of it, rather it should be seen as maintaining a deferred reprocessing option allowing for change in long term energy policy and spent fuel management.

3   Interim Storage Systems

Vault storage systems have been developed to store large quantities of irradiated fuel and storage of vitrified high level wastes. They can be designed to receive fuel in a transport cask or fuel which is previously sealed into a storage canister. The fuel or fuel canister is normally stored vertically. Vault stores can be equipped to operate independently of an associated reactor pool virtually as process plants in their own right. Vaults can have a high initial capital cost, and have a degree of permanence which is not always preferred by the owner/operator. The modularity of other systems enables expenditure profiles extended over long term periods.

Wet storage is a well proven technology given its existing role in reactors and for the cooling of fuel prior to reprocessing. However, moving to interim storage and away from prompt reprocessing puts pressure on pool capacities. Whilst pool re-racking continues to be a cost effective option for the short term, finite pool capacity remains the ultimate constraint. Further pool storage for very extended periods is also likely to be both expensive and detrimental to fuel cladding.

The most successful interim dry storage solution to date rests with storage only casks or dual purpose storage and transport casks. Frequently, the dual purpose cask design is derived from an existing single purpose cask. Casks are manufactured from steel - the more expensive method - or from concrete, the lowest cost option.

4   BNFL and the Interim Storage Market

The demand for interim storage is growing. With this in mind and with a view to ensuring that BNFL's portfolio of fuel cycle products and services continues to meet customer needs BNFL has now entered the interim storage market. BNFL has entered into a close partnership agreement with Sierra Nuclear Corporation (SNC) of the United States, an already well established and successful organisation in the interim storage market with a number of contracts in the USA and the Ukraine.

SNC is responsible for designing the ventilated concrete cask system for the safe storage of irradiated fuel for up to 50 years which is in use in the United States. Together BNFL and SNC have modified the original system to provide a dual-purpose, transport and storage design called TranStor™. This comprises the already proven concrete cask and transfer cask designs with the newly developed canister suitable for transport, and storage within the new TranStor™ shipping cask.

The TranStor™ system has been designed to meet the requirements of the United States Code of Federal Regulations and the International Atomic Energy Agency (IAEA) Safety Series 6 shipping regulations. The TranStor™ system is fully compatible with BNFL's world-wide shipping system and can interface with its reprocessing facilities at Sellafield. To date the BNFL/SNC systems have been fully developed for PWR and BWR fuels, and conceptual designs prepared to confirm the feasibility for VVER and RMBK designs.

BNFL's principal offering in the interim fuel storage market is the TranStor™ concrete cask system. In addition, BNFL also provides casks for the transport and storage of vitrified high level waste (VHLW) which has been developed from the BNFL vitrified waste transport flask. This is covered later in the paper.

5   Overview of TranStor™

The storage system is designed for safe long-term storage of spent nuclear fuel. It is designed to survive normal, off-normal, and postulated accident conditions without an unacceptable release of radioactive material or excessive radiation exposure to workers or members of the general public. Storage components are designed and fabricated in accordance with recognised codes and standards that provide ample safety margin.

Design features that have been incorporated in the Storage system to assure safe long-term fuel storage include:

- Leak-tight/multi-pass welds on canister structural lid, shield lid, shell, and bottom plate.
- Thick lids and walls to minimise radiation exposure to public and site personnel.
- Design of canister body and internals to withstand a postulated drop accident during storage or transportation.
- Design of concrete cask to protect canister from postulated environmental events.
- Use of coatings to minimise contamination of the transfer cask by fuel pond water.

Operation of the TranStor™ system requires the spent fuel to be loaded into a specially designed sealed canister. Fuel loading takes place within the fuel pool, during which time the canister is held within a shielded steel transfer cask. Upon completion of fuel loading the canister is lidded and whilst still in the transfer cask, the canister is removed from the fuel pool. The canister shielding lid is then welded shut, the canister is vacuum dried, back filled with helium and a second structural lid is welded in place. If fuel is to be shipped off site the canister is then transferred directly from the transfer cask to a shipping cask.

The primary components of the TranStor™ system are:

the sealed canister for storage and off site transport which provides the containment for the spent fuel;

the concrete cask, which provides physical protection and shielding for the canister during storage and is used for short journeys on the reactor site from the reactor building to the storage pad;

the transfer cask which provides shielding during fuel loading and transfer operations into and out of the concrete cask or the shipping cask;

the shipping cask which is used to transport a loaded canister off site to a central storage site or on completion of the storage period, to a final disposal site or a reprocessing facility.

The TranStor™ canister is shown in Figure 1. It is double seal welded and contains an array of steel storage sleeves, inside a circular steel shell. The base of the canister is a large steel disk welded to the canister shell. The canister has two upper lids, the shield lid and the structural lid. These are placed on the canister after it is loaded with fuel and they are welded to the canister shell.

The transfer cask is a metal cask with a steel, lead and neutron shield composite radial wall. See Figure 2. This transfer cask serves to move the canister to and from the storage pool, the concrete cask and the transport cask. It has thick steel doors in its base to accommodate vertical loading.

The concrete cask is a large, circular cylinder of concrete with an internal steel liner and is shown in Figure 3. The steel liner serves to reduce the concrete wall thickness needed for shielding and provides a form during concrete pouring. The concrete cask provides for shielding of the fuel inside the canister and provides the natural circulation air flow path for cooling the canister. The cooling air passes up through air ducts running the length of the concrete casks. The use of vertical storage greatly enhances the decay heat removal capability of the system compared to horizontal systems. The concrete cask can be easily transported around the reactor site.

The shipping cask is shown in Figure 4 and is a composite walled cask designed to transport the loaded canister anywhere in the world. The transport cask and its payload of the loaded canister are designed to meet US Code of Federal Regulations and IAEA Safety Series Six requirements. Hence, the system is compatible with shipping fuel to BNFL's facility for further storage and reprocessing or to other storage facilities world-wide, this keeping reactor operators options open.

The spent fuel is stored within the concrete casks on a simple reinforced concrete pad. The pad has a compact layout minimising the land required. No building or other covering is essential although one can be constructed at minimum cost if necessary to enclose the casks and minimise visual impact. Storage of the casks is undertaken vertically to minimise the storage area required and associated costs. The storage pad facility may include security fencing, lighting, temperature instrumentation and air flow and radiation monitors depending upon local requirements.

The TranStor™ system base design can store zircaloy clad fuel. Work is underway for generic licensing applications for other types of fuel. The system can accommodate a heat load of up to 26kw for storage and 24kw for transport. This can be met by various combinations of initial enrichment, burn up, assembly weight and decay time.

The TranStor™ system's components utilise the experience generated from the operating experience gained at reactor sites in the USA combined with the extensive experience of BNFL shipping cask operation. The system is based on the Ventilated Storage Cask already licensed by NRC and submissions for the TranStor™ storage and shipping licenses have been made. No difficulty is anticipated in demonstrating compliance with other country's national regulations which are based on IAEA transport regulations and have much in common with US NRC requirements.

6   Description of the TranStor™ Canister

The primary functions of the canister are:

- Provide containment and confinement for the radioactive materials during normal storage and credible or postulated accidents
- Provide criticality control and structural support for spent fuel in the presence of a moderator
- Provide adequate heat transfer to prevent fuel temperature from exceeding allowable limits under design and abnormal conditions
- Provide a unit which when placed inside a shipping cask, may be transported to IAEA standards.

The canister is a transportable canister consisting of an outer shell assembly, a shield lid, a structural lid, and an internal canister assembly. It is designed to store 24 PWR fuel assemblies (BWR-61, VVER 1000-24, VVER 440-85, RMBK - 162 assemblies) and has provisions for accommodating four (4) failed fuel cans (in place of four intact fuel assemblies). The shell provides the pressure boundary and is designed to withstand postulated accidents without loss of containment integrity. The canister shell is made out of stainless steel, ensuring no unacceptable corrosion over the nominal cask lifetime of 50 years.

The internal surfaces are coated to prevent detrimental effects on the fuel pond water chemistry during the time the canister is in the pond for loading. The coating is radiation resistant and can withstand temperatures up to 650°C.

The canister shell and the bottom plate are fabricated from Type 304L stainless steel. The canister bottom plate is welded to the shell in the fabrication shop.

The internal canister assembly is fabricated from carbon steel plates formed into an array of 24 square storage cells. Each cell holds one PWR fuel assembly. The four outer corner cells are slightly larger to allow placement of a failed fuel can. The canister contains flux traps formed by structural tubes and BORAL sheets placed within the cell. Therefore, it is criticality safe for unburned fuel and nonborated water. Structural tubes also provide support for the sleeves and fuel during a postulated drop accident, thus satisfying the IAEA Regulations for the transport of nuclear fuel.

The canister closure is accomplished by the use of a two component system. The first component is a shield lid made of Type 304 steel plates. It is placed on the canister after the fuel has been loaded. Two penetrations through this shield lid are provided for draining, vacuum drying, and back filling the canister with helium. The drain penetrations utilise nominal pipe thread fittings on the top and bottom. A drain pipe is screwed into the bottom of the shield lid prior to lowering the lid into the pond. The pipe extends to the bottom of the canister to facilitate water removal. This penetration is also used to blow air into the canister in order to remove residual water. The other fitting is a quick disconnect fitting used for vacuum drying and helium back filling. The canister is tested for leak tightness (1 x 10^{-4} skd cc/sec at 0.5 Atmospheres pressure differential) prior to final sealing of penetration cover plates.

The other component of the canister closure is a structural lid. This lid is a Type 304L plate that has a penetration for access to the shield lid fittings. This penetration is sealed via multiple welds once the helium backfill process has been completed. Both shield and structural lids are welded to the canister shell after the fuel is inserted.

7   Description of TranStor™ Transfer Cask

The primary functions of the transfer cask are:-

- Provide radiological shielding for personnel during canister sealing and   transfer
- Provide canister surface protection from pond contamination
- Provide lifting capabilities for the canister during fuel loading and transfer.

The transfer cask is used for canister transfer from the spent fuel pond to the concrete cask prior to storage and from the concrete cask to the Shipping Cask for off-site shipping. It consists of a cylinder with a steel-lead-neutron shield-steel sandwich wall that reduces the dose rate at its surface. The top cover of the transfer cask extends over the canister to prevent it from being inadvertently lifted out of the top of the cask during transfer from the Spent Fuel Pond to the concrete cask. The bottom of the transfer cask has retractable shield doors to allow lowering of the canister into the concrete cask. These shield doors also reduce the dose rate to workers as the transfer cask is lifted to the top of the concrete cask for canister placement. Hydraulic pistons are used to retract the doors for the canister transfer.

The transfer cask is lifted from above by the cask lifting yoke via two trunnions located on the outer shell approximately three feet from the top of the transfer cask. The lifting yoke is fabricated from high strength carbon steel and is used for all cask handling operations in the Fuel Building. The trunnions are solid steel and extend radially from the cask body. Each trunnion is welded to the inner and outer steel shells of the transfer cask wall with full penetration circumferential welds. The two trunnions are capable of accommodating the combined weight of the transfer cask and a fully loaded wet canister while meeting the requirements of NUR 0612. The transfer cask is fabricated in accordance with ANSI N14.6 requirements and its lifting components are tested to 150% of their maximum design load. The yoke provided with the transfer cask is used to interface with the existing Fuel Building crane and has provisions to prevent contamination of the crane hook and block.

8   Description of the TranStor™ Ventilated Concrete Cask

The primary functions of the concrete cask are:

- Protect the canister from weather and postulated environmental events such as rains, winds, tornado missiles
- Provide stability during the site design earthquake
- Provide adequate convective cooling for the canister
- Provide adequate shielding (together with the canister) to meet 10 CFR 72 requirements.

The concrete cask provides structural support, shielding, and natural convection cooling for the canister. The canister is stored in the central steel lined cavity of the concrete cask. The cask is ventilated by internal air flow paths which allow the decay heat from the fuel to be removed by natural circulation around the metal canister wall. Air flow paths are formed by the skid channels at the bottom (air entrance), the air inlet ducts, the gap between the canister exterior and the concrete cask interior, and the air outlet ducts. The air inlet and outlet vents are steel lined penetrations that take non-planar paths to minimise radiation streaming. Side surface radiation dose rates are limited by the thick steel and concrete walls of the cask.

The concrete cask is a reinforced concrete cylinder. The concrete contains Type II Portland Cement. Outer and inner re-bar cages are formed by vertical hook bars and horizontal ring bars. The concrete mix has been specifically selected to assure strength and long life at the elevated temperatures expected during normal operations (38 ~ 95°C) and the higher short-term temperatures that could potentially occur during off-normal and accident conditions. The Type II Portland Cement was selected to match the aggregate's and its carrier's thermal expansion coefficients and because it requires a low water/cement ratio.

The internal cavity of the concrete cask is formed by an A-36 steel liner and bottom plate. The steel and concrete walls of the cask are designed to minimise side surface radiation dose rates.

The air flow path is formed by the openings at the bottom (air entrance), the air inlet ducts, the gap between the canister exterior and the concrete cask interior, and the air outlet ducts at the top. The air inlet and outlet vents are steel-lined penetrations that take non-planar paths to minimise radiation streaming. A shield ring is provided over the canister-liner annulus to reduce the dose rate at the top of the cask.

The cask lid is fabricated from an A-36 plate which provides a cover and seal to protect the canister from the environment and postulated tornado missiles. The lid is bolted in place and is provided with a tamper indicator. It also provides additional shielding to reduce the skyshine radiation.

The bottom of the concrete cask is covered with a steel plate which minimises loss of cask concrete during a bottom drop accident. The concrete cask has reinforced chamfered corners at the top and bottom to minimise damage during handling.

The cask is constructed by pouring concrete between a re-usable form and the inner metal liner. The reinforcing bars and air flow embedments are installed and tied prior to pouring.

The thick walled concrete cask provides good radiation protection and its vertical orientation provides efficient heat removal.

Since the cask does not become contaminated during service, it can be disposed of in a normal landfill or can be reused as a low level waste disposal package. In either case disposal is simple and inexpensive. The height of the cask is approximately 5 metres, dependent on the fuel type being stored, resulting in minimal visual impact and a low centre of gravity for stable transfer operations.

To date the concrete casks have been constructed generally at the reactor site, either at the storage pad location or inside a redundant building or other enclosure. The steel liner forms the inner wall with the rebar cage built around it. The air inlet and outlet channels are welded to the inner shell and tied to the outer rebar frame. The concrete casks also include temperature monitoring instrumentation located in the air outlets.

Historically local construction companies with nuclear power plant experience have been employed to fabricate the concrete casks to specification at the site. However, the casks can be constructed off-site at a local contractors facility and transported to the plant.

9   TranStor™ Shipping Cask Design

An illustration of the shipping cask is shown in Figure 4. The shipping cask consists of an inner and outer steel shell assembly, lead gamma shielding between the inner and outer shells, neutron shielding outside the outer shell, and a stainless steel jacket around the neutron shielding material. The inner and outer shells are welded to top and bottom inner and outer. Neutron shielding material outside the structural shell is a solid synthetic polymer. The shipping cask inner shell rings, top forging, bottom inner forging and the inner lid establish the cavity to accommodate the canister.

The shipping cask has a fully recessed closure that is protected from side impacts. The closure contains a vent port and captured primary and secondary seals. The vent and drain port design includes a bolted cover that retains primary and secondary gaskets. All ports are located under the impact limiters for protection from postulated fire and impact events. The port covers are of small diameter to ensure the cask meets the IAEA punch test.

Lifting capability for the packages is provided by four lifting trunnions. The trunnions extend radially from the cask body at 90-degree intervals and made of solid steel. Each trunnion is bolted to the inner and outer steel shells of the cask wall.

Energy-absorbing impact limiters are installed over each end of the cask, to provide protection in the event of an accident during transportation. The impact limiters are made of Redwood/Balsa Wood encased in Stainless Steel and dissipate any impact force(s) that may be applied to the cask during an accident. The impact limiters are removable and are bolted over upper and lower ends of the cask.

10 Licensing

SNC and its personnel have been working with the US Nuclear Regulatory Commission (NRC) on the licensing of dry fuel storage systems since the inception on the concepts in the early 1980's.

BNFL itself operates the worlds largest nuclear shipping fleet (150 casks) and has been involved in designing and licensing shipping casks from the early 1970's.

The original ventilated storage cask system (VSC) was awarded an NRC license in 1993. The new transportable version of this system TranStor™ has been jointly designed by BNFL and SNC and is currently being assessed for approval by the NRC. Final approval of generic NRC licenses for PWR and BWR fuels is expected in early 1997.

11 Vitrified High Level Waste Storage and Transport Flask (VISTA)

General Description

The assembled cask consists of four main components, the body, the primary lid, the secondary lids and the internal support structure. The body is manufactured from carbon steel forgings with the base welded to the body shell. Cooling fins are welded to the cask's surface. Neutron shielding material is contained in sealed compartments at the fin roots and the cask base.

The lids are manufactured from a carbon steel forging and incorporate a double metallic seal system. The lids, which are equipped with a lifting pintle, are bolted to the cask body. Neutron shielding material is enclosed on the upper surface of the primary lid, while on the underside, annular spaces locate the top ends of the residue containers and provide impact protection. A sealed valve gives access through the lid to the cask cavity to allow gas sampling to be carried out.

The internal support structure is built up from a number of aluminium segments and provides accommodation for 28 vitrified residue containers.

The cask is equipped with a pair of trunnions at each end of the body, which are used for lifting purposes and for supporting and securing the flask during transport. When prepared for transport the cask has steel clad, wood filled shock absorbers attached to each end. The shock absorbers provide mechanical protection for the cask.

Internal Support Structure

The internal support structure consists of aluminium segments assembled to form circular compartments. One compartment runs centrally along the cask's axis with the other positioned radially around it. The segments are clamped against the cask wall to provide stability and optimum heat transmission. The system of clamps and axial spacers between the segments allows differential thermal expansion to take place without causing unacceptable distortion of the support structure. Each stack of aluminium segments is located on a full length key, screwed to the cask's cavity wall. The segment clamps are attached to the key and are designed to be operated by a remotely controlled tightening system.

Radiation Shielding

The major part of the gamma shielding is provided by the thick steel walls and ends of the flask. The fin structures and other fittings will make only a small contribution to the gamma shielding performance. The uniform nature of the vitrified residue means that the axial dose rate profile is also uniform over the length of the cask cavity.

Vitrified waste contains neutron emitting isotopes and to attenuate the neutron radiation, a layer of neutron shield material is installed in compartments at the fin roots and at the ends of the flask. This material provides neutron shielding and ensures that the neutron dose rate is relatively uniform over the flask's surface.

Heat Transfer Systems

The flask operates dry under all circumstances and, consequently, heat transfer from the contents to the flask body is effected principally by a combination of conduction and radiation. The heat radiates from the residue containers into the aluminium segments which form the internal support structure. The clamping system which secures the support structure ensures that close contact is maintained with the flask wall. The heat is then conducted through the flask wall into the circumferential cooling fins, from where it is dissipated to the surrounding atmosphere.

Thermal Performance Design

The flask has a capacity for 28 vitrified residue containers with a total decay heat output of 56 kW with the heat output from any individual container not exceeding 2.5kW. The flasks' heat transfer systems are designed to ensure the residue is kept well below the minimum phase transformation temperature of the vitrified product.

Impact Performance Design

Each end of the flask is fitted with a steel clad, wood filled shock absorber during transport. These shock absorbers provide impact protection for the ends of the flask. The system of fins, together with the enclosed neutron shielding compartments, provides protection in the case of the flask being subjected to a side impact. On the underside of the flask lid, flared spacers locate the lid ends of the waste containers and protect the neck of the container against damage caused by movement under normal operations.

Sealing Systems

Each flask penetration is sealed by a double metallic seal system equipped with an interseal test point. Each pair of seals is tested by pressurising the interspace with air and observing the pressure change.

Operational Features

The flask has a pair of trunnions attached to each end of the body. These turnings have a larger inner journal which supports the flask during transport, while the smaller outer journals are used in conjunction with a lifting beam for flask handing purposes. The flask primary lid incorporates a filling valve which gives direct access to the flask's interior for evacuation and helium has filling prior to transport. The secondary lid provides facilities for checking the seal of the primary lid.

12   Conclusions

- The TranStor™ system embodies the proven storage technology from the SNC Ventilated Storage Cask with BNFL transport design and operating experience. This offers an economic and versatile solution for spent fuel interim storage.

- The BNFL VISTA Vitrified High Level Waste storage and transport flask provides an economic alternative to construction of vaults where only small quantities of High Level Waste are to be stored.

- Both of these developments complement BNFL's existing spent fuel management services to the nuclear industry.

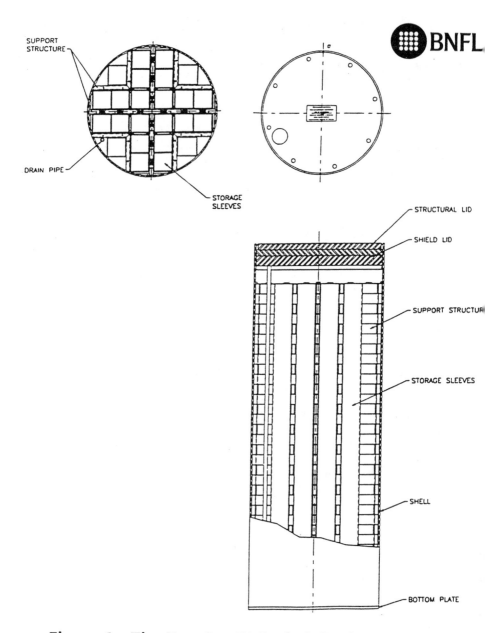

SUPPORT STRUCTURE

DRAIN PIPE

STORAGE SLEEVES

BNFL

STRUCTURAL LID

SHIELD LID

SUPPORT STRUCTURE

STORAGE SLEEVES

SHELL

BOTTOM PLATE

Figure 1 - The TranStor™ Sealed Canister

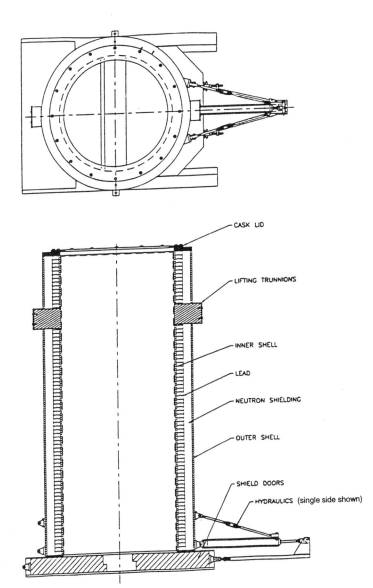

Figure 2 - The TranStor™ Transfer Cask

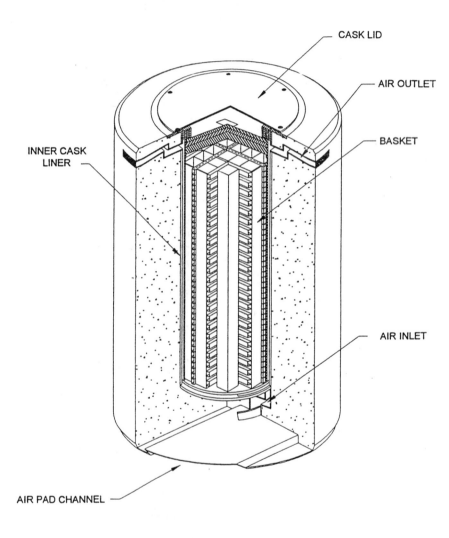

CASK LID

AIR OUTLET

BASKET

INNER CASK
LINER

AIR INLET

AIR PAD CHANNEL

Figure 3 - The TranStor™ Concrete Cask

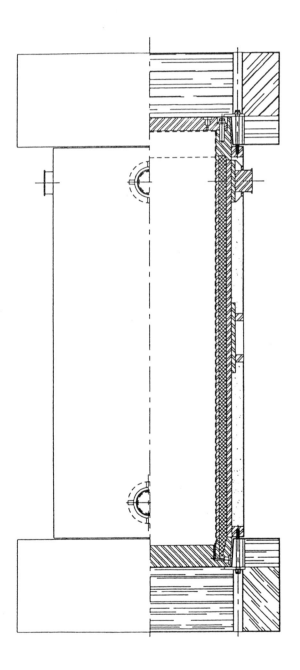

Figure 4 - The Transtor Shipping Cask

w intermediate level waste (ILW) recovery and packaging :ilities for UKAEA Harwell

ALL BSc, CEng and **D B SYME** PhD, MInstNucMaterMan
Technology plc, Didcot, Oxfordshire, UK

Synopsis

New facilities are being commissioned at Harwell which will enable the retrieval of α β/γ Intermediate Level Waste (ILW) which is currently held underground in existing stores in deteriorating waste cans.

The facilities provide for processing and packing of the waste into 500 litre stainless steel drums which are then stored in an integral Vault prior to ultimate Nirex disposal.

Paramount in the design of these facilities is the avoidance of criticality and other hazards, and a computer based SCADA system controls operation of the plant which incorporates a state of the art non-destructive assay (NDA) system to analyse the waste and hence optimise packaging.

INTRODUCTION

Within the United Kingdom, the UKAEA strategy for the management of alpha contaminated beta/gamma ILW is to recover current stocks from their present location and process them to give immobilised products that are acceptable to Nirex, the UK Agency responsible for disposal. The ILW storage facilities at Harwell have been designed to implement this strategy and comprise:

- Improvement of the superstructure, containment and floor shielding of the present stores to improve radiological safety during waste retrieval operations. Provision of mobile, shielded retrieval equipment for the recovery of wastes from their present unsatisfactory storage.

- Construction of a Vault Store with input Head End Cells (HECs) that will process and load the waste into stainless steel 500 litre drums and, eventually, immobilise the waste ready for Nirex disposal.

- Provision of equipment for Low Level Waste (LLW) and Contact Handleable Intermediate Level Waste (CHILW) diversion for separate processing.

The project will allow the retrieval of existing solid ILW, which is currently held underground in deteriorating mild steel cans. It will transfer this and new arisings to the HECs for processing and onward for dry storage in a heavily shielded vault. Much of the plant is of novel design and full size mock-ups were constructed at Harwell for testing and developing operational and maintenance procedures. The new facilities are subject to licensing by the UK Nuclear Installations Inspectorate (NII) who have maintained close liaison as the project has progressed. To allow for changes of waste processing strategy during operations the facility has been designed with flexibility as a key consideration.

This has been a UKAEA project funded under the DRAWMOPS programme of the Department of Trade & Industry. AEA Technology plc have been responsible for the design and project management of the project together with supply of the specialised assay equipment.

DESCRIPTION OF THE FACILITIES

Introduction

The local site plan showing the layout of the new facilities including the personnel building is shown in Fig. 1. The existing Alpha, Beta/Gamma stores are linked to the new facility (462.27) by the Interim Store (462.26).

Retrieval Equipment

There are some eight thousand existing cans of intermediate level α β/γ solid waste stored in Harwell buildings 462.2 and 462.9. The waste cans, which are in a variety of sizes from ten to fifty litres nominal capacity, are stored in over one thousand vertical storage tubes below the ground level up to four and a half metres deep. Each storage tube is topped with a concrete filled shield plug. The number of waste cans per tube varies between four and eleven. The Retrieval Machine (RM), effectively a mobile shielded cell, has been designed to move over each storage tube, emptying them one at a time.

The RM comprises two modules: a lower shielded and upper module, which together weigh approximately 90 tonnes (see Fig.2). Each module is moved and repositioned in the building using a purpose made handling frame mounted on floor rails. Each module has a stainless steel containment surrounded by 250 millimetres of lead shielding. The containments mean that any damaged waste cans encountered can be retrieved safely without any release to the surrounding environment. As elsewhere (in building 462.27) the retrieval machine uses an engineered doubled-lidded posting system at the cell boundaries which places retrieved waste in a sealed transfer drum for a safe onward processing.

The upper containment module contains two transfer ports: one is aligned over the storage tube the other interfaces with a transfer drum. Also provided are two master slave manipulators, an in-cell hoist with a waste can retrieval grab(s) and a CCTV system used for viewing down the storage tubes.

For waste recovery the lower shielded module is firstly positioned over the selected storage tube, a sealed connection is made between its containment and the top of the storage tube.

Next, the upper shielded module and gamma gate are located onto the lower module and all control and ventilation connections are made to the machine. A shielded transfer flask (containing an empty transfer drum) is placed on the gamma gate.

Waste retrieval then follows a cycle of: removing the storage tube shield plug, lowering a grab from the cell hoist to engage a waste can, raising it into the upper module and placing it in a transfer drum which is then sealed and returned into the flask. The flask is unloaded at the HECs and the empty transfer drum returned to the RM where the process is repeated. When the storage tube is empty the shield plug is replaced and the RM is moved to the next selected storage tube.

Head End Cell Suite

The HEC Suite (Fig 3) comprises a row of concrete shielded cells connected by intercell and undercell transfer systems. Each cell is provided with a lead glass viewing window, master slave manipulators and roof mounted lighting. There are four compartments each 2.7m x 1.9m x 2.6m lined with stainless steel and with shielding walls 1500mm thick.

The sequence of operations is shown in Fig 4. Cans from the input facilities are first fed to the Input Cell via two double lidded ports in the cell floor and a horizontal posting port in the end wall. Here they are assigned a unique bar code label before transferring to the assay cell.

The assay cell is provided with an active/passive neutron interrogator and a high resolution segmented gamma spectrometer. The neutron interrogator is used to provide confirmation of the existing record of fissile content of the waste can. The fissile content of each waste can is tracked through the facility as part of the criticality safety system providing guidance for packing into the larger containers. The inventory of gamma-emitting radionuclides provided by the gamma spectrometer is also used to confirm the can contents against the existing plant records, but, in addition, it also provides an up to date assessment of the can contents. This information will be required when the waste is taken to its final repository. Based on the assay results the can is assigned to an ILW or LLW/CHILW stream. It then moves into the Packing Cell.

The Packing Cell contains instrumented storage positions for a number of waste cans. Output from sensors on these positions is used to determine the most effective packing order for cans being loaded into new stainless steel 500 litre drums using the in cell hoist and grab. Waste cans designated as LLW or CHILW are overpacked in 500 litre drums and diverted from the vault store for processing elsewhere. The Packing Cell also contains equipment to open the waste cans and tip their contents onto an inspection tray for transfer to a 500 litre drum. The empty waste cans are placed in a hydraulic press and crushed to form "pucks" which are retained in the cell until they too can be diverted as part of the LLW/CHILW stream.

The lidding cell is separately contained from the other cells, and from the under-cell transfer area. It is used to attach a secondary bolted lid and lifting feature on to the waste drums prior to storage. The lidding cell also contains a swabbing facility and monitoring equipment. These are used to check seals and lids to ensure that the drum is contamination free prior to its transfer to the Vault Store.

Vault Store

The Vault Store is an above ground, seismically qualified structure with outer shield walls of reinforced concrete 1400mm thick. It is capable of storing some 2,700 Waste Drums. Internal walls divide it into eight storage bays of which seven will be used for drum storage. The eighth will be used for temporary storage allowing movement and retrieval of drums to satisfy safety requirements. A filtered ventilation system with separate inlet and extract systems is provided to deal with an external temperature range of -10 to +40° C. The fire resistance is estimated to be 4 hours.

All drum handling operations are carried using a remotely operated overhead travelling crane. A shielded maintenance area for this is located adjacent to the vault.

Control Room and Supervisory Computer System

The plant control room provides operators with a clear view of the operational face of the HEC. In addition to the waste management computer system it incorporates the following safety and control equipment:

(i)   Hardwired cell line mimics
(ii)   Ventilation mimic and controls
(iii)   Fire panel
(iv)   Scram initiator
(v)   Health Physics mimic or indicators
(vi)   Intercom and Public Address system
(vii)   Vault store crane controls
(viii)   Cell line, under-cell tunnels and transfer area closed circuit TV systems
(ix)   Limited remote ventilation indication from the plant complex.

Bar codes are read automatically by a remote camera system. This confirms can identity and initiates the transfer of specific can-related information to the assay systems which then provide results that are used by the Supervisory Computer System to confirm that criticality safety conditions are met, to recommend a packing plan and to control the packing sequence of the 500 litre Vault Store Drum.

SAFETY CONSIDERATIONS

Licencing and Safety Protocols

The Waste Storage and Shielding Facility is subject to the statutory provisions for licencing nuclear installations in the U.K. The granting of a site licence to operate requires the satisfaction of the NII by a series of formal Safety Case submissions and protocols.
A detailed analysis of the safety aspects of the plant has been carried out covering all operations from the entry of the flask containing the waste into the facility, to the storage of the waste drums and the provisions for recovery of waste from the store. The safety of future expansions of the plant (these could include waste immobilisation and the final unloading of the Vault Store) will be considered when the decision is made to immobilise the waste form.

Hazard Assessment

Normal Operation

Normal radiation levels in the plant have been established according to the ALARP principle. Assessments show that shielding design, facility containment and method of operation will allow the plant to be operated well within the basic criteria that stipulates the maximum permissible annual dose to the operators (5mSv average as a limit, 15mSv as the maximum to any individual) and to members of the public (50μSv maximum to any individual).

Fault Conditions

Potential faults leading to the release of activity or the exposure of radioactive materials were identified systematically using established Hazard and Operability (HAZOP) techniques. Faults were assessed probabilistically using fault tree analysis, and risks to workers and the general public were shown to comply with the risk criteria established by HM Nuclear Installations Inspectorate (NII). The facility has been designed to expedite the recovery from fault situations.

Risks to operators are dominated by direct exposure hazards at gamma gates and shield doors and are minimised by the use of interlocks to prevent hazardous operations.

The only other significant hazard from internally initiated events was identified as a criticality incident. However using pessimistic assumptions about critical mass configurations the assessed frequency of a criticality incident was shown to be well within the NII target frequency criteria. The assessment took account of the reliability of equipment and procedures provided to identify, measure and track fissile materials, and of the fact that assembling a critical mass would require the combination of fissile materials from separate sources.

Externally initiated events have also been assessed, an aircraft crash being by far the most significant event when the risk and frequency criteria are still met. The requirement to seismically qualify the facility in terms of shielding but not containment in the event of a design base earthquake was also identified.

Plant Life and Decommissioning

The nominal life of the facility is 25 years (50 years for the storage vault itself) and as outlined above, it will initially be used to store raw waste and then possibly conditioned waste pending its disposal in a waste repository. Due to the nature of the operations the store itself will remain relatively uncontaminated as each drum placed in the store will be clean. All HECs, where there is potential for internal contamination, are lined with stainless steel to allow easy cleaning. Items difficult to decontaminate can be fed into the waste stream. The design of the facility has taken ease of decommissioning into account and therefore decommissioning of the facility is not expected to lead to unacceptable hazards.

Management Operations

The facility will operate under a written 'Authority to Operate'. All activities in the facility will be carried out in accordance with written procedures and will be subject to an integrated quality assurance programme. Waste movements will be centrally planned and a record will be kept of the content of each waste package and its location in the facility. Health Physics monitoring will be carried out according to written procedures to ensure safe operation of the facility. The operations carried out at the facility will be reviewed and reported on annually to the NII in order to support the justification for its continued operation.

CONTROL AND INSTRUMENTATION PHILOSOPHY FOR SAFETY

Cell Line Controls and Monitors

The computer systems employed to control the Vault Store and Head End Cell can be partitioned into three main interconnected areas namely, the process control system which is responsible for overall control of operations, the assay system that conducts the gamma and neutron analyses of the waste cans and the database system that registers, for permanent record, information concerning the can and 500 litre vault drum contents. In addition, there is an expert system, based on a stand-alone PC which is used to provide advice to the control room operator in reconciling any major discrepancies between the assay results and the historical waste can records.

Process Control

Process control is performed by a network of PLCs and a central control room PC. Each cell has an associated PLC and a cell workstation and, when required for operator interaction with the process, a simple on-line monitor with keypad. The workstation resides alongside the other hardwired controls in a command console below the cell face window.

The control room computer incorporates colour monitors with dynamic mimic displays covering all plant operations. Control over specific operations is authorised from the control room only but for others it may also be local to the plant.

Where risk reduction is practicable, fault tolerance from erroneous PLC commands is provided by autonomous hardwired protection circuits. The inputs to the protection system are fully segregated from the process control system.

Criticality Control Philosophy

The philosophy of criticality control within the facility is to limit the mass of fissile material to safe levels at predetermined positions/areas within the facility. For example, within the vault store, control is achieved by limiting the mass of fissile material within each storage drum such that the large array of drums in the store will be safe. Theoretical safe limits were initially established by an extensive set of Monte Carlo simulations. Practical theoretical limits are taken to be three standard deviations less than this.

To facilitate the maintenance of criticality safety, the intention is to:

a) identify the cans and record their progress through the HEC suite.

b) quantify the fissile arisings at the facility, primarily using the existing historical records of waste can contents, with direct confirmation by measuring the fissile content of waste cans after entry into the Head End suite

c) record the fissile inventory and location of storage drums being sent to the Vault Store.

To ensure criticality safety the fissile content of each can is assessed from the historical can inventory and its comparison with the measured fissile content. Appropriate interlocks are provided to limit the fissile mass in each area of the facility. It should be noted that no specific control requirements are currently envisaged over the quantities of moderator present within the facility. The fissile inventory within the facility is confined to waste cans and storage drums. In the event that one of these is found to exceed the set criteria, it is removed from the facility for appropriate processing elsewhere.

Plant Database

The Plant Database records exist on two separate relational databases. The Waste Can Database holds information on the identification, contents and origin of each of the processed cans. The Vault Store Drum Database contains similar information for each 500 litre drum and includes the identity of waste cans in the drum and its location in the Vault Store. The two databases reside on a single PC.

ASSAY INSTRUMENTATION AND MEASUREMENTS

Two assay instruments perform measurements on the waste cans before they are combined and the waste packaged in 500 litre drums:

The Gamma Spectrometer

Inside the assay cell, the waste can being measured is placed on an elevating turntable assembly immediately in front of the collimator; the can is rotated and raised past the collimator aperture so that its entire surface can be scanned by the germanium detector to determine its gamma-emitter content.

The Gamma Spectrometer measurement is performed on one horizontal slice (or segment) of the can at a time; a gamma-ray spectrum is collected for each segment with the can rotating at 30 rpm to even out the response to any radial inhomogeneities in the waste density or activity distribution. Each segment spectrum is individually analysed using a data analysis package to determine the radionuclides present and their activities from the characteristic gamma-ray peaks present in the spectrum. The results are corrected for the effects of gamma-ray attenuation within the waste matrix and the process is repeated for each segment of the can and the results summed to produce a total inventory for the waste can.

The Neutron Interrogator

Neutron interrogation involves placing a waste can inside a 'neutron well counter' and measuring the neutrons emitted from the fissile material present in two modes:

- **Passive Mode,** in which one counts the neutrons produced by the spontaneous fission of the even isotopes of plutonium (^{238}Pu, ^{240}Pu and ^{242}Pu), using the coincidence counting technique.

- **Active Mode,** in which one exposes the waste cans to neutrons from a sealed ^{252}Cf source to induce fission in the fissile material present (mainly ^{235}U and ^{239}Pu), removing the source to a shielded 'home' position and counting the delayed neutrons produced in the short lived fission products produced, ie the 'Californium Shuffler' technique.

The results of the fissile and spontaneous fission measurements will generally enable the masses of uranium and plutonium to be separately estimated.

Data Reporting and Reconciliation with Plant Records

After both the gamma spectrometer and neutron interrogator assay measurements are complete, the assay results are displayed to the control room operator on the screen of the Assay PC along with the original plant record for the waste can. The control room operator must then decide whether the assay results and the plant records are consistent with each other (taking into account the quoted measurement uncertainties). If the assay results and plant records do not agree, he must decide whether:

- the plant records are incorrect

- the can has been wrongly identified so that the plant records are for the wrong can

- there is some factor that may be affecting the assay results to cause a measurement bias over and above the estimated measurement uncertainties (even after the correction factors detailed above have been taken into account) and, if so, which data should be entered into to the waste can and Vault Store databases.

The consistency of both the fissile and gamma radionuclide inventory measurements with the plant records is clearly an important factor in making such a decision. In order to facilitate this process, an expert system has been developed into which both sets of assay results are entered following an assay measurement, along with the key items from the plant records for the waste can.

The expert system knowledge base consists of a set of rules concerning the inter-relationships between these various parameters, knowledge about the factors affecting the measurement (matrix effects, homogeneity, limits of detection) as well as physics-based rules (e.g. radionuclide half-lives to allow it to take into account decay of short-lived nuclides since the original records were produced). It uses these to draw a set of conclusions from the input data. From this and the responses given by the control room operator to further questions prompted by the particular lines of reasoning, it offers advice on the likely cause of any discrepancy.

TESTING AND COMMISSIONING

Three stages cover the testing and proving of the facility in a logical progression towards active operations. Biological shielding tests were performed when construction of the cells was completed.

Setting to Work and Testing

This is an in situ extension of the shop testing which the various components have undergone previously. The aim is to check the operational and functional integrity of the installed system including the physical set ups and calibrations to enable testing of combinational interlocks and sequential operations to be performed. The assay system was set to work assaying simulated waste cans containing known quantities of (sealed) gamma radionuclide and fissile samples. General de-bugging is undertaken to ensure a trouble free operation whilst the facility is freely accessible.

Inactive Commissioning

Activities during this stage are being directed at:

- ensuring that the facility can be maintained by remote operations

- undertaking specific tests identified in the Safety Commissioning Schedule developed from the HAZOP and ongoing Fault Tree Analysis

- simulating full process operation using dummy cans and data etc. aimed at demonstrating procedures and equipment reliability and Operator Training.

Active Commissioning

This will involve repeating the process operations using initially known, well quantified active waste and extending this subsequently to a limited quantity of material from the standard waste stream. A representative range of mixed waste types will be retrieved for this purpose. Further checks will be carried out on the procedures, safety rules etc. which apply.

STATUS

Construction of the facility began in 1991and Inactive Commissioning and operator training are now nearing completion.

ACKNOWLEDGEMENTS

This work has been funded under the DRAWMS/DRAWMOPS programme of the UK Department of Trade and Industry (DTI). The results of this work form part of the UK Government programme on decommissioning and radioactive waste management, but do not necessarily represent Government policy.

We would like to express our thanks to the DRAWMOPS Directorate of UKAEA for their support during the project and for permission to publish this material.

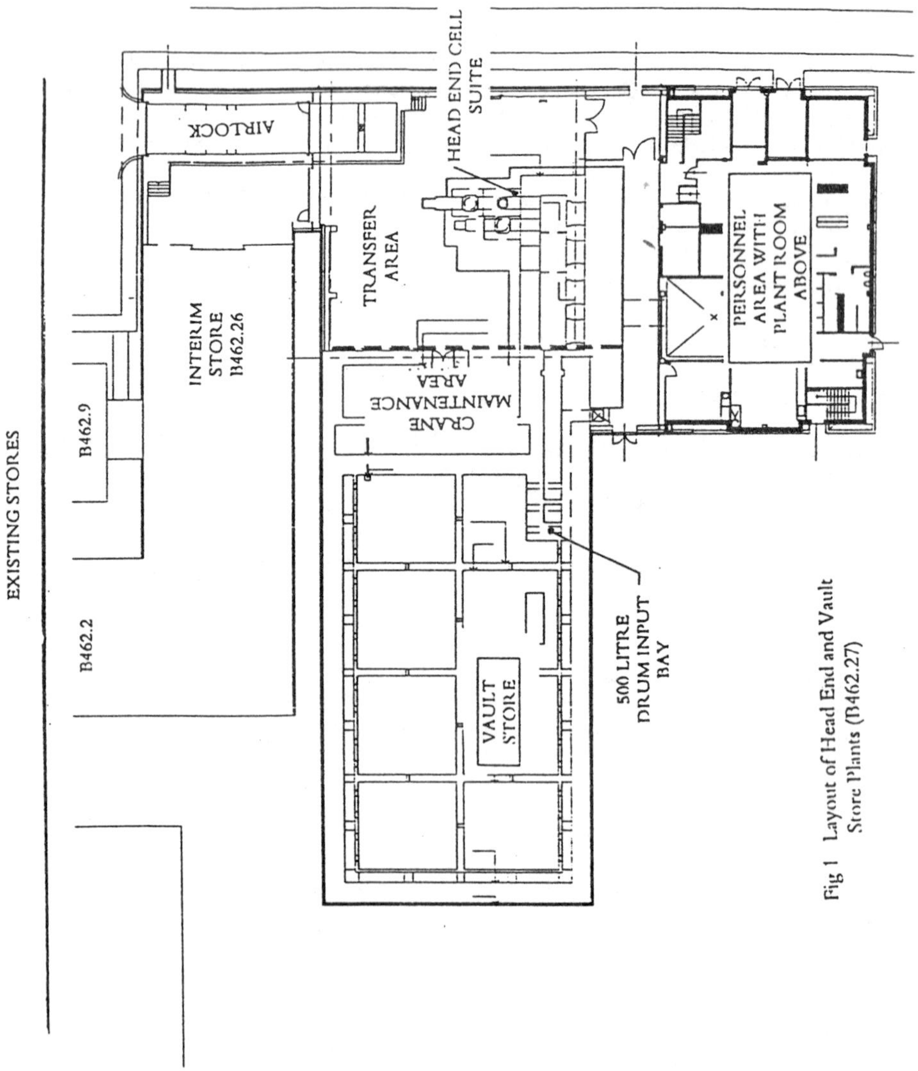

EXISTING STORES

B462.2

B462.9

INTERIM STORE B462.26

AIRLOCK

HEAD END CELL SUITE

TRANSFER AREA

CRANE MAINTENANCE AREA

VAULT STORE

PERSONNEL AREA WITH PLANT ROOM ABOVE

500 LITRE DRUM INPUT BAY

Fig 1 Layout of Head End and Vault Store Plants (B462.27)

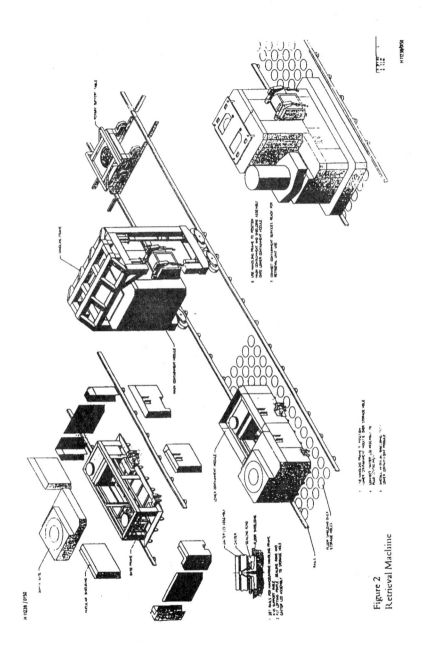

Figure 2
Retrieval Machine

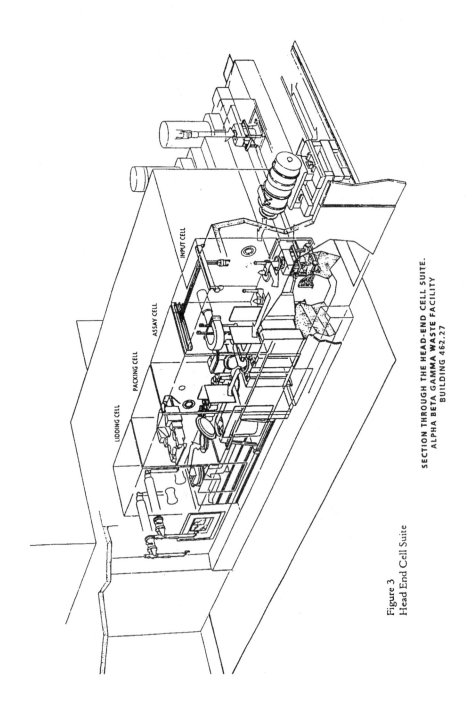

LIDDING CELL

PACKING CELL

ASSAY CELL

INPUT CELL

SECTION THROUGH THE HEAD-END CELL SUITE.
ALPHA BETA GAMMA WASTE FACILITY
BUILDING 462.27

Figure 3
Head End Cell Suite

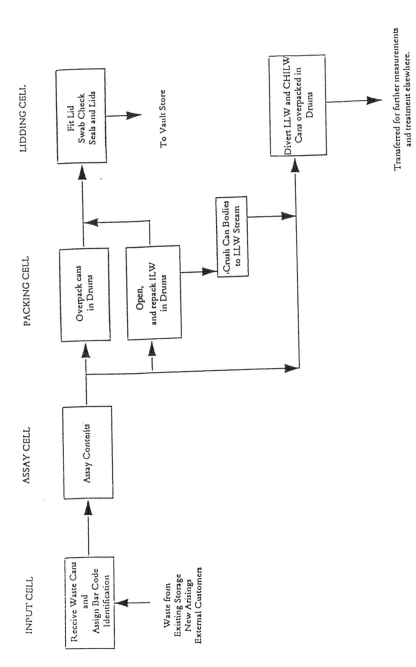

Fig 4   Head End Cells – Sequence of Operations

erational experience with storage of vitrified highly
ive waste at BNFL Sellafield

HEATLEY BEng, CEng, MIEE
Nuclear Fuels plc, Sellafield, UK

BNFL's strategy for the management of Highly Active liquor fission product waste is one of Vitrification and then temporary storage at Sellafield. To meet this requirement the Vitrification Plants on the site comprise the Waste Vitrification Plant, the Vitrified Product Store (VPS) and the VPS Export Facility. This paper will describe the principles and operational experience from the storage of Vitrified Product Containers in the VPS, and summarise the future storage philosophy.

1 PLANT DESIGN AND OPERATIONAL PHILOSOPHY

1.1 History of the Vitrification Process

The reprocessing of spent nuclear fuel at Sellafield gives rise to a solution, containing 99% of the dissolved fission products with impurities, known as Highly Active Liquid Waste (HAL). The impurities are primarily from the spent fuel cladding materials, which include traces of unseparated plutonium (Pu), uranium (U) and most of the transuranic elements. The HAL is evaporated and stored as an aqueous nitric acid solution in high integrity stainless steel tanks (HASTs), that require cooling systems to remove fission product decay heat, and agitation systems to maintain solids in suspension. Storage experience in HASTs has been excellent, though continual surveillance over long periods is costly.

Vitrification, based on waste solidification and immobilisation of the radioactivity, is a desirable alternative. Such a system is insensitive to loss of services and does not require the same degree of supervision, giving advantages in terms of safety, economy, handling and stability for storage, transportation and ultimate disposal.

Vitrification development began in the 1950's, leading to a contract being signed by BNFL for the French Company SGN to provide the design for the Waste Vitrification Plant (WVP). Construction began in 1984, and the first container transferred to the VPS in October 1990.

1.2 The Vitrification Process

WVP consists of a HAL storage and distribution cell, two parallel Vitrification lines (consisting of a Vitrification Cell and a Pouring Cell), a container decontamination cell and a container monitoring and control cell, (Fig 1). The HAL is converted by evaporation and calcination into highly active solid, mixed with glass solids, and heated in an electric induction furnace to produce a molten vitrified waste. This is poured into 400kg capacity stainless steel containers and a lid welded onto each prior to decontamination, monitoring and transfer to the VPS. Container transfer is by means of specially designed, dedicated and heavily shielded flasks.

The VPS function is to store containers until natural radioactive decay reduces their heat emission. Foreign generated waste will be returned to the originator via the Export Facility.

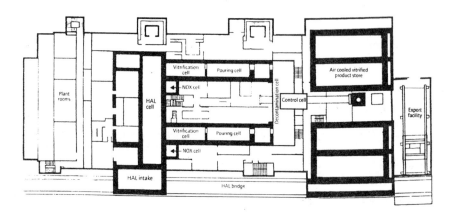

Fig 1 Plan of the Waste Vitrification Plant

2   LAYOUT AND LOADING OF THE VPS

The 7960 container capacity is housed in four natural convection air-cooled compartments, each with 199 channels, and 10 spaces per channel (Fig 2). Each channel comprises a thimble tube (in which the containers reside), a thimble cap (to shield the top of each tube), and floor plug (to maintain containment and shielding between the channels and the operating area).

Negligible contamination levels and operator doses within the VPS have been experienced, since rigorous decontamination and monitoring processes are applied to the product containers, prior to despatch from the WVP.

Each compartment will be filled in turn in accordance with a pre-determined loading pattern, designed to evenly distribute the heat loading across the compartment. A designated storage channel is prepared to receive a container by positioning a mobile gamma gate over the channel, and the Charge Floor Flasks are used flask out the floor plug, flask in a charge chute (to bridge the gap between the top of the thimble tube and the charge floor) and flask out the thimble cap. The mobile gamma gate remains in position to maintain radiological shielding during all flasking operations.

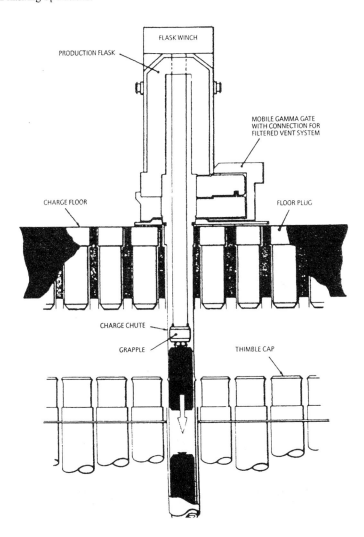

Fig 2 Cross Section of the Vitrified Product Store

The decontaminated vitrified product container is loaded into a heavily shielded product flask in the WVP Control Cell, and transported to the VPS flask handling area by an electrically driven rail bogie. The flask is lifted, by the 50 tonne overhead crane onto the mobile gamma gate, above the open channel. The gamma gate is opened and the container lowered into the channel using the product flask winch (Fig 3). The procedure is repeated until the channel has been filled, and then the channel is closed off.

The equipment available within the VPS allows two channels to be open at the same time, thus enabling continuous loading operations, which has proved to be the most efficient loading strategy with respect to manpower requirements.

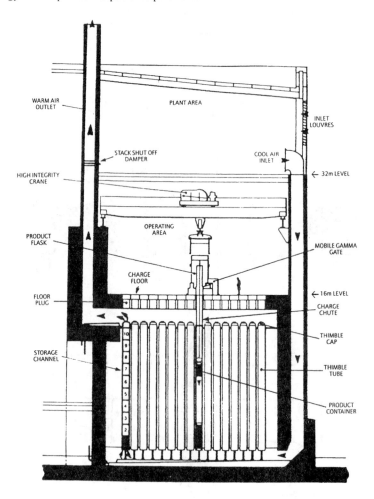

Fig 3 Flasking Arrangement for loading product containers into the a VPS channel

3   OPERATIONAL SAFETY CRITERIA

The Nuclear Installations Inspectorate, the main Regulatory Authority, require completion and approval of a fully developed Safety Case (fdSC) prior to operation and to be fully reviewed every two years. The fdSC addresses all potential hazards within the building. It concluded that one mandatory 'Operating Rule' is necessary, and identified safety significant equipment that must be periodically demonstrated to operate correctly ('Safety Mechanisms'). The designed levels of interlock protection are defined by a company standard.

3.1 Operating Rules

The one Operating Rule for the VPS concerns the Bulk Outlet Temperature of the storage compartments. In order to prevent a breakdown of the glass and product waste (devitrification which occurs if the container contents exceed 500°C) within the containers, the compartment bulk outlet temperature should be kept below 170°C. If this temperature is exceeded it must be reduced to an acceptable level within 24 hours.

The following plant parameters have been designed to meet the temperature restrictions :

i)   Maximum container decay heat   = 2.5kW
ii) Maximum thimble tube rating (10 containers) = 20.0kW
iii) Maximum compartment rating   = 3.8MW

To ensure the operating rule is not breached, all critical store temperatures are monitored at the building control room, via a dedicated computer system. Alarm levels are set to indicate any significant increases, with initiation of remedial action at 140°C. The outlet temperature of any compartment can be controlled by manual adjustment of the ventilation system.

The average decay heat for containers produced from reprocessed MAGNOX fuel is 1.2kW, and the Bulk Outlet Temperature is maintained approximately at 60°C. At the end of 1995, over 1000 containers had been transported to VPS for storage. Extrapolation of the channel temperature data shows that the bulk air temperature in the first compartment will be maintained within the operating parameters as the compartment is filled to capacity.

3.2 Safety Mechanisms

There are three identified 'Safety Mechanisms' on the VPS equipment, failure of any one of which could lead to an incident in which operators could be exposed to high direct radiation :

i) The gamma gate mechanical locking system, which prevents the gamma gate door opening without a flask in position on it.

ii) The hardwired container detection maintenance gamma gate interlock system, which prevents a product container being lowered into an unshielded maintenance area.

iii) The product flask trunnion blind Castell key protection systems, which prevent the flask being lifted off a gamma gate when the shielding door has not been locked in position.

Each of these Safety Mechanisms has been reviewed against site safety criteria, and engineered to provide adequate protection against loss of shielding incidents. There have been no failures of these Safety Mechanisms, though operational checks are made monthly.

3.3 Storage Vault Structure

Each storage compartment is a rectangular vault, 18m long, 7.5m wide and 16m high, enclosed within concrete biological shielding approximately 1.5m thick. Each vault is made inaccessible before loading may commence.

The operating floor forms the roof of the storage compartment. It is a monolithic structure of steel and concrete, penetrated by the channel access holes, providing biological shielding, while allowing thermal expansion of the floor. A layer of aluminium honeycomb is built into the upper layer of the operating floor, which would absorb the impact of an accidentally dropped flask, without imposing excessive loads into the structure.

The thimble tubes are secured to the vault floor. Each tube has a shock absorber at the base, which support the weight of 10 product containers without collapsing, but to collapse and absorb the energy of an accidentally dropped container, without damage to either the container or the thimble tube. The tubes are also located by lower, mid and upper diaphragm plates and structural steelwork.

Containers within the thimble tubes are supported in such a way as to protect against potential seismic activity, with aluminium honeycomb buffers at the corners of the vault floor to provide seismic restraint. Since all adjacent buildings and pipe bridges have also been built to protect against such activity, there is no seismic threat to the structural integrity.

An outer stainless steel tube encloses the inner carbon steel thimble tube of each channel. The lower portion of each outer tube is coated with aluminium metal spray. It is this area that will be subject to the greatest corrosion attack by moist air from the natural ventilation. When the compartments are being filled, there will be sufficient heating from the containers to dry the ventilation air, and prevent corrosion. For the compartments that will be idle for a number of years following the start up of the store, a dehumidification system has been installed.

The structural concrete must be maintained below 100°C to prevent premature ageing. Insulation is provided to reduce the heating effects from the product containers. Temperatures in the concrete and steelwork structure are monitored to ensure conditions are continually acceptable. Sample specimens are positioned around the building for removal and analysis, to assess the long term effects of the heat and radiation on the building structure.

Environmental conditions within the store have proved to be satisfactory and maintainable.

4 MODIFICATIONS TO CONTAINER TRANSPORT EQUIPMENT

Safety analysis of the operation of the VPS during 1992, which incorporated updated safety criteria, required an upgrade to the then existing Safety Mechanisms.

The project was structured to allow the VPS operations to continue through the modifications, which were undertaken over a 12 month period.

A typical example of the modifications required is the addition of a diverse interlock to detect the presence of a replaced floor plug under a mobile gamma gate. A mechanical lock was installed to consolidate hardwired and software detection methods that were already employed.

Additional work was undertaken to update and improve the flasking system's operability. This work included installing multisafe amplifiers to detect any faults in the umbilical cables used to connect all charge floor equipment to the PLC control system. The modified system acts to prevent operations continuing if a failure is detected.

All the modifications undertaken during the 1992 project have proved successful.

5   VAULT COOLING

The fission products and actinides, encapsulated in the glass can release up to 2.5kW of decay heating in each container. This heat must be removed to prevent the de-vitrification of the glass, and to prevent an excessive temperature build up within the concrete walls of the shielded vaults.

The process of cooling the vitrified product containers is achieved by a natural convection air flow that passes over the outer surface of the thimble tubes within which the product containers are vertically stacked (Fig 4). Air cooling was selected for this process because it offered the most dependable medium for heat transfer over a long period. Natural convection cooling was prefered to forced convection, because the process is inherently simple, which leads to a design which occupies the smallest site area, provides the lowest unit storage cost and requires minimum surveillance.

The cooling air does not contact the product container surfaces, and is discharged at low level. The cooling air does not contribute to the site active effluent discharge levels which has been confirmed by Beta-in-air stack monitors, temperature probes and flowmeters installed on each compartment stack.

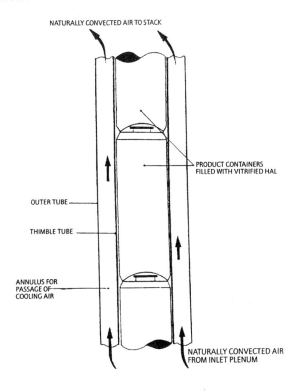

Fig 4 Detail of Storage Tube showing cooling arrangement

Cooling air enters the building via the large banks of louvers that are situated at the 32m level on three sides of the building. The air passes into the compartments via inlet ducts, which are constructed of concrete, and are an integral part of the compartment structure. The air then passes around the bottom of the thimble tubes in the lower plenum, from where it is drawn up over the thimble tubes and passes into the upper plenum. From the upper plenum the warm air rises up the outlet stack and is discharged to atmosphere. Test work showed, and it has been demonstrated on the plant, that this system produces a positive air flow through the store under all weather conditions.

6   INSTRUMENTATION

6.1 Ventilation

Any instrumentation that is deemed to have any safety implications, such as Beta activity in the stack, stack air flow rates, stack and inlet air temperatures, is powered by a guaranteed non-interruptible supply and is hardwired to the control room to provide essential information should the control and monitoring computer fail, or the building lose its normal power supply.

The cooling air system is under manual control and there are no power requirements.

6.2 Temperature Monitoring

Thermocouples are positioned on each storage channel, which indicate any high temperatures generated by individual channels. They are also positioned at points in the vault steelwork and concrete structure. High temperatures are logged and alarmed to ensure remedial action is initiated if required.

6.3 Computer System

The VPS computer monitors the plant status and provides a container inventory system. The computer logs the operating parameters of the store, such as temperatures and ventilation information, and provides displays, reports and alarm functions. The inventory feature records data on container location and loading date, to allow verification of container identity and provide traceability. The computer does not perform any plant control function.

6.4 Flasking Equipment

The flasks and gamma gates have in-built interlocks to prevent loss of shielding incidents, such as a shield door opening inadvertently. These interlocks combine mechanical, hardwire and software layers of protection. All signals showing the correct, or incorrect, operation of the equipment, are fed to the controlling programmable logic controllers. The interlock functions allowing continued operations are generated by proximity and reed switches.

The integrity of equipment detection devices is checked continually during operation.

Gamma monitors are also installed on the Product Flasks to detect the presence of a container. This signal, which is displayed on the side of the flask, is also used to provide a diverse interlock function for the maintenance facility, to prevent exposure to a container when direct maintenance of the flask grapples is required. Such an important interlock has four layers of interlock provided via the control instrumentation, and has been upgraded since operations commenced.

7 FUTURE STORAGE PHILOSOPHY

7.1 Future Investment

Operation of the plant has been characterised by good performance from the glass making process, with the consistent manufacture of in-specification product containers. Early experience demonstrated that product container throughput was restricted by extended periods of downtime between production campaigns in the WVP. An improvement programme resulted in 330 containers being transported to the VPS during 1994/95. Operation to date has shown that the VPS equipment is satisfactorily reliable at the current production rates.

However, the operation of the Export Facility (due to begin operations in 1997) and the 3rd WVP Line (due to begin production in 2000), require improved operational efficiency. To that end, a study is being undertaken by the Operational Research Department to model the VPS, and identify areas for operational improvements, and if the purchase of new equipment is required to maintain operations for the designed 50 year plant lifetime.

7.2 VPS Export Facility (VPSEF)

The store cannot provide indefinite storage facilities for the site vitrified waste production. Therefore containers must either be moved to a final repository or in the care of the overseas customers be returned to the country of origin. The VPSEF provides the container destorage route. The VPSEF is assessed in the existing Vitrification Plant fdSC, and so conforms to all plant safety standards.

The export process provides containment and shielding from the containers (Fig 5). The VPSEF has an inspection cell that receives containers from a VPS storage channel, via the VPS flasking equipment.

Inspected containers are lowered by the in-cell crane into a multi-chamber export flask. When the flask has been filled, with up to 28 containers, it is transported, via bogie and the overhead 130 tonne crane, onto a rail transporter for shipment off site.

7.2 Conclusion

The operation of the VPS has proved to be successful with surveillance and manpower requirements being provided by the WVP personnel This method of supervision has demonstrated that the design requirement to minimise the necessary human intervention, has been effective.

The stability of the Product Containers has allowed future operations to be predicted, and safety has been maintained as the highest priority. Management of reviews for the equipment and operational requirements within the VPS will allow the continued demonstration of the advantages of the Vitrification Process.

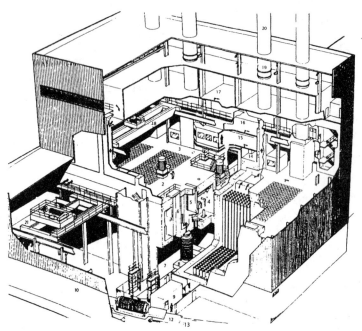

1	Flask Entry	11	Export Flask
2	Product Flask	12	Export Flask Transporter
3	Charge Floor	13	Railway Track and Road
4	VPSEF Inspection Cell	14	Flasking Equipment
5	VPSEF Hoist Park	15	VPS Control Room
6	Drop Well	16	Switchgear Room
7	Export Flask Handling Area	17	Building Ventilation Room
8	Export Flask Bogie	18	50 tonne VPS Crane
9	Change Room	19	Outlet Stack Damper
10	Export Flask Building Exit	20	Outlet Stack
		21	VPSEF 130 tonne crane

Fig 5 Pictorial arrangement of the Vitrified Product Store and Export Facility

REFERENCES

(1)   MILLINGTON D. Progress with highly active waste vitrification at BNFL Sellafield, The Nuclear Engineer, 1995, Volume 36, pages 42-45.

Packaging and storage of plutonium in BNFL's THORP complex

PARKES Bsc, BA, Phd
British Nuclear Fuels plc, Sellafield, UK

1.   SYNOPSIS

BNFL's THORP plant began active commissioning in 1994. One of the products from this plant is plutonium for recycle back into reactors. The THORP Plutonium Finishing Line has been extensively developed for remote operation to accommodate the high temperatures and radiation levels associated with high burn up derived plutonium.   An all steel packaging system has been developed to afford good thermal conductivity and excellent containment to meet international test criteria.   The store has been designed to afford long term storage against an array of accident scenarios.   The paper describes; operation of the packaging and storage facility, safety aspects, design considerations, quality assurance and control, handling into storage and to the associated Mox plant, regulations, safeguards, and operational experience.

2.   INTRODUCTION

When civil reprocessing of irradiated fuel began, storage of plutonium was seen as an interim measure until the material was used to fuel the coming generation of fast neutron breeder reactors.   As fossil fuel and uranium prices fell and costs for a breeder system rose, it was seen that utilisation of plutonium on commercial scale in fast reactors was receding into the next century.   Large scale commercial capacity to utilise plutonium as Mox fuel in thermal reactors is now under installation which will result in a net decline in plutonium stocks world-wide.   In the interim, systems for indefinite safe storage of plutonium have been developed to meet the challenge.

3.   PACKAGE DESIGN

Fuel from Magnox reactors only achieves a low burn-up of under 5 GWd/tU, and the plutonium extracted from it during reprocessing is relatively easy to handle in terms of heat from Pu238 and radiation from Pu241 (Table 1).   Packaging of plutonium from the Magnox finishing line involves filling a screw top aluminium can and 'bagging' this out in a polythene intermediate package into a stainless steel outer can which is resistance welded to give total containment.

Table 1. Change in isotopic content (%) and heat output of plutonium with increased burn-up of reprocessed fuel.

Burn-up level in GWd/t(U)	W/kg	Pu238	Pu239	Pu240	Pu241	Pu242
<1	2	0.05	93.6	6.0	0.4	<0.05
5 (Magnox)	3	0.2	69	25	5	1
30 (LWR)	10	1	60	22	13	4
60 (LWR)	30	4	48	24	12	12

Fuel from Advanced Gas Cooled reactors (AGRs) and Light Water reactors (LWRs) achieves significantly higher burn-up and the plutonium extracted from it has a different isotopic content (Table 1) which results in higher heat and radiation output. The heat output from oxide derived plutonium would almost immediately fail a polythene outer. Instead, a stainless steel intermediate bagging procedure was developed.. The additional steel reduces the radiation level to some extent, but the levels are still high (Table 2). The gamma radiation increases with time due to the build up of Am241 from Pu241.

Table 2. Dose rates from a 'reference' THORP plutonium package.

Time after separation	Can	Approximate dose rate at can surface (mSv/h)	
		Neutron	Gamma
28 days	Inner	20	10
	Outer	10	5
20 years	Inner	20	150
	Outer	10	25

The THORP package underwent several revisions during design and testing. These were mainly carried out to;
- Improve the reliability of the sealing and un-sealing of the screw top
- Simplify mechanical handling
- Optimise the thickness and dimensions to increase the internal and external failure pressure while allowing a good weld and cut
- Improve the resistance to bursting during drop tests onto targets
- Aid location of the components prior to welding and cutting.

Proof testing of package designs involves destructive testing by pressurisation of dummy loaded containers under normal and fault conditions, i.e. after drop tests, at normal and elevated temperatures. The packages withstand nearly 100 atmospheres of internal pressure (Table 3) and even higher external pressure. The final product is leak tested to demonstrate compliance with international transport requirements.

Table 3. Internal pressure test results for THORP plutonium package

Temperature (°C)	Failure pressure (MPa)
20	>9
200	>6
400	>4.5

The three close fitting stainless steel layers in the final package, shown in Figure 1, afford good heat dissipation as well as resistance to physical damage. This is desirable as high burn-up oxide derived plutonium can have a heat output of over 100 W for a 5 kg package (Table 1) which can result in oxide temperatures in excess of 200 °C at the can wall and 500 °C at the centre line.

4. PREPARATION FOR STORAGE

The principle method of finishing plutonium, that is converting from an aqueous nitrate stream to an oxide, is oxalate precipitation. The conditions for this precipitation are chosen to minimise the production of fine powder which could become airborne during a loss of containment, while keeping the particle size small enough to allow subsequent solid state reaction required to homogenise the product with uranium for fuel production. The effect of process conditions on the birth to growth rates during crystallisation are captured in a computer code which can be used to predict the crystal size range of the product. A range between 20 and 100 micron is desired.

The particle size influences the specific surface area of the powder, but this can be further tuned using the temperature during thermal decomposition of the oxalate to the oxide. Too low a surface area renders the oxide un-reactive, while too high a surface area could lead to excess adsorbed gas which has the propensity to cause pressurisation of the storage container. A surface area of 5-15 m2/g is suitable for Mox fuel production while meeting the storage specification of < 0.25 w/o loss on heating for the THORP store. This is achieved for product in the standard particle size range using a calcination temperature of around 550 °C. The effect of moisture on potential pressurisation are modelled in a computer code.

5. THE PACKAGING PROCESS

Due to the high heat and radiation output from the plutonium, all operations are remotely viewed and operated, the bulk of operations being under a computer sequence.

The gloveboxes in which operations are carried out are constructed of seamless stainless steel to minimise the potential for dust to become entrained. Packaging is carried out under an inert atmosphere to minimise adsorbed gas and moisture on the powder product.

Components are transferred into and out of the packaging line using bagless transfer systems which minimise the generation of plutonium contaminated waste while maintaining containment of plutonium.

The calcined plutonium oxide is remotely transferred from a hopper into a screw top stainless steel container on a weigh cell. After re-placing the lid, the unique can number is read using a bar code reader and transferred, along with all other data, to the central control and information computer.

The inner package is placed inside an intermediate can sitting in a bagless transfer port. A bung is placed behind the inner in the neck of the intermediate can. The intermediate can is rotated in the transfer port and a laser is used to weld a seal between the bung and the intermediate can. After welding, the laser is re-focused and used to cut along the centre line of the weld. The laser technique is the most accurate method of cutting along the centre-line of the weld and avoids the generation of swarf. The laser is remote from the cell with the beam transferred through an optical window to minimise equipment in the glovebox. The bung has a 'blow hole' that prevents pressurisation in the bung which would result in blowing out of the weld.

The welded intermediate is extracted from the adjacent 'clean' glovebox, leaving the off-cut bung to seal the port. The bung is displaced into the active glovebox by the next intermediate can.

After remote number reading again, the intermediate is inserted into an outer can sitting in a port to the adjacent cell. The outer is moved to a new position where it is evacuated and back filled with helium for leak detection purposes before a lid is inserted. The outer is repositioned and rotated in a resistance welder to seal the lid. The weld area is automatically swabbed to ensure containment of radioactivity.

The can is moved to a vacuum chamber for leak detection to ensure that it complies with international transport requirements. The can is then moved to a buffer store before transfer to the main store.

6.   STORAGE

The THORP plutonium product store is constructed from thick re-inforced concrete to afford physical protection. Cans are stored in cells within horizontal re-entrant tubes which provide an additional level of containment between the product and the environment. Any loss of containment would be detected by sniffer tubes, seen on the cell face in Figure 2.

The Magnox stores rely upon the use of manually operated shield blocks for the loading and unloading of PuO_2 packages. Use of this arrangement for oxide derived plutonium would result in un-acceptable doses. Containers are loaded and unloaded from the tubes in the THORP store by a fully automated loading and retrieval aisle stacker. This system can only hold one plutonium package at any time to ensure criticality safety.

The cells are tall enough to create a chimney effect which would ensure adequate removal of the decay heat from the plutonium. The cells are cooled by forced air, however, in order to keep the concrete cool and extend the lifetime of the store. Redundancy in all power and control distribution lines is provided. The concrete structure has been tested to ensure that any possible temperature rises from a full store which has suffered a total loss of forced ventilation will not cause structural damage. Criticality safety within the cells is assured by geometry.

The use of remote systems not only reduces dose uptake to operators, it also helps increase the security of the store by allowing strict control of access through a secure perimeter. The manner and material of construction, and installation and operation of appropriate surveillance equipment also serve to maintain the security of the store. As with all other PuO_2 facilities at Sellafield, the store is subject to verification by EURATOM Safeguards, with a continuous presence of EURATOM inspectors on site. Design features to assist in the inspection and verification of store contents include secure seals for storage channels and vault, weighing and NDA equipment to assess the contents of storage containers, and the ability to inspect containers within channels. Also available are the continuously recorded surveillance data. As with other BNFL facilities the store is open to inspection by the IAEA.

Purpose designed and built containers, which meet IAEA standards, enable safe transport of THORP packages of PuO_2 to other facilities or to foreign customers. The containers hold up to eight THORP packages and have been thoroughly tested against impact, fire and immersion to ensure that containment will not be breached in the event of an accident.

When the large scale Sellafield Mox Plant comes on-line at the turn or the century, plutonium packages will be transferred directly across from the store to the adjacent plant via an engineered link.

7.   COMPARISON TO OTHER SYSTEMS

Savannah River Laboratory in the US are currently developing a THORP type system, i.e. one which uses a steel bagging system in a bagless transfer port, following publication of an early article by BNFL (1). This system is manually operated and uses a resistance weld followed by mechanical cutting, rather than laser welding and cutting. The welder must be removed and the cutter located accurately along the centre line of the weld. All the mechanical components are located in the glovebox.

The BNFL system utilises a sealed outer can but the intermediate package has a powder filter. Thus, should the package pressurise and the outer fail, the inner will vent through the filter and the powder will be contained, i.e. the system is fail-safe. The Russian packaging system has a ceramic filter in the outer containment. The French package does not incorporate a filter.

There have been many incidents of loss of containment of packages in the US (2,3), primarily due to organic materials being packaged with plutonium. Recent US specifications for passivation and packaging are rigorous and stipulate very high calcination temperatures (4).

The US and the Soviet Union have historically built their plutonium storage facilities underground as protection against military attack. Commercially operated and licensed stores in the UK and France use above ground storage.

8. OPERATIONAL EXPERIENCE

Being the custodian of the world's largest stockpile of civil plutonium, BNFL has extensive operational experience in packaging and storage of plutonium.

The several thousand packages produced from the Magnox finishing lines are essentially the same as those from THORP, since no credit is taken in the qualification tests for the inner and intermediate packages. These are routinely inspected and have been observed over a long time period. The contents of packages containing Magnox derived plutonium, however, generate much lower heat outputs as well as much lower radiation fields.

THORP is in the middle of its commissioning (at the end of 1995) and plutonium packaging represents the final stage of the process. Commissioning has been carried out in stage-wise campaigns and three of these campaigns had progressed to the finishing line by the end of 1995, representing several tens of packages successfully completed.

9. PLANS FOR THE FUTURE

THORP's plutonium packaging and storage system represents the state of the art and moves are afoot to market these proven and commercially available systems in the US to assist with the problems left over by their military campaigns of plutonium production.

Advances in laser technology have come about since the system was procured and characterised and efforts are underway to replace the laser power pack with a more compact unit which will offer reduced cycle times and higher availability.

Product from THORP will be transferred through an engineered link from the plutonium store to the Sellafield Mox Plant when it comes on line in 1997. A new generation of packaging is already under development and qualification which will minimise the amount of scrap packaging on material destined for short term use rather than for long term storage.

10. CONCLUSIONS

BNFL's plutonium packaging and storage system is state of the art and represents several generations of evolutionary development to cope with arduous conditions for material storage.

11. REFERENCES

1   W BAXTER, 'Plutonium finishing and product packaging in THORP', Atom 397, November 1989, p32.

2   'Assessment of plutonium storage safety issues at DoE facilities', DoE/DP-0123T, January 1994.

3   'Plutonium working group report on environmental, safety and health vulnerabilities associated with the department's plutonium storage', US DoE/EH-0415, September 1994.

4   'Criteria for safe storage of plutonium metals and oxides', US DoE standard 3013, December 1994.

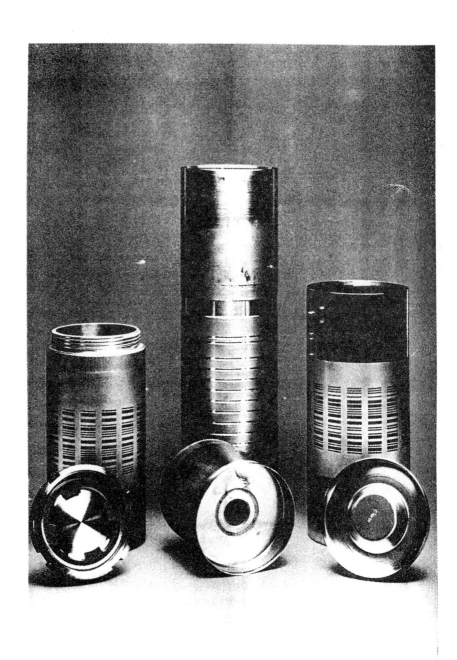

Figure 1: The inner, intermediate, and outer which form the Thorp product can

© IMechE 1996 C512

Figure 2: Photograph inside the Thorp plutonium store

512/038/96

1e spent fuel pool re-racking option for Central and 1stern Europe

ANDERSON BS, MS, MAMNucSoc, INMM, NEI
1 Associates, Connecticut, USA

1. INTRODUCTION

In the past few years, every country in this region has had to consider a new approach to spent fuel management. Most of the countries now plan to store spent fuel for many years followed by direct disposal. Except for Russia and the Czech Republic, the countries in this region are not presently considering independent central storage either. The near term requirement is for spent fuel storage at-site for a long time.

Dry storage technologies have been considered and in some cases are being implemented. But such technologies have to be adapted in a period of political and regulatory uncertainty. This takes time and money which is not readily available in the region today. A near term alternative is spent fuel pool re-racking brought about by advances in re-racking design and analysis.

In the west, conversion from low to maximum density fuel racks has been made possible by several advances in analysis methods as well as changes in licensing requirements, such as:

- Use of neutron poison materials
- Free standing rack designs
- Credit for fuel burnup in criticality analysis
- Regionalized spent fuel pools
- Use of hydrodynamic coupling in dynamic analysis
- Three dimensional multi-rack seismic analysis

Nearly 150 re-racking campaigns have been performed in the US alone. Many of these involved second or third campaigns at the same plant in order to accommodate evolving US spent fuel management policy. These same techniques may now be applied to meet the changing requirements in the Central and Eastern European region.

There are three major reactor types in the region; VVER-440, VVER-1000 and RBMK. In order to simplify the discussion, this report will focus on the VVER-1000 design. However, the concepts and principles can be extended to the other fuel types as well.

2.   GENERAL DISCUSSION

The considerations that influence a utility's decision to re-rack, as well as those that dictate the ultimate benefits of re-racking, include:

- Pool area available for rack placement
- Minimum rack center-to-center spacing achievable
- Pool structural adequacy
- Fuel pool cooling system limitations and/or maximum heat removal capacities
- Dose rate minimizations in areas adjacent to the pool
- Spent-fuel handling equipment travel limitations
- Overall program costs and schedule.

In addition, utilities need to anticipate the increased demands caused by the use of high performance fresh fuel:

- Higher enrichments (e.g. $3.2 \rightarrow 5.0$ W/O)
- Higher burnup (e.g. $33 \rightarrow 60$ GWD/MT)

3.   MAXIMUM DENSITY VVER RACK DESIGN

3.1   Rack Design

The objective of maximum density storage rack design is to meet all licensing requirements with pool spacing that approximates the same assembly pitch in the reactor core.

Figure 1 shows the basic VVER-1000 storage cell and module design. The storage cell is approximately 258 mm on pitch---only slightly larger than the fuel assembly itself. It is made of 11 gage (3.2 mm), 316 stainless steel with a poison insert wrapper on the outside walls of each cell. The inserts are about 2 mm thick with an areal poison density of .01-.03 gm/cm^2. In the US the poison insert is either Boraflex, a silicone material infused with natural boride carbide powder, or Boral -- which is boride carbide mixed in aluminum. Borated stainless steel has been used once as an insert but licensing of borated stainless steel as a structural material has never been achieved in the US (although it is used in Europe).

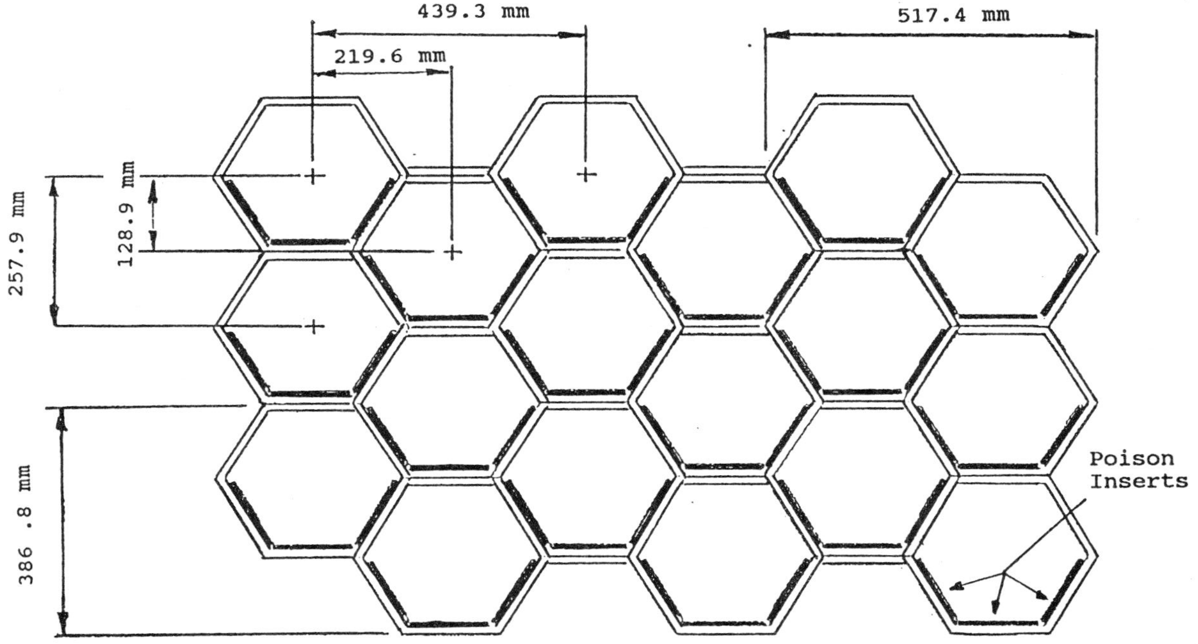

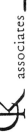

Poison
Inserts

517.4 mm

439.3 mm

219.6 mm

128.9 mm

257.9 mm

386.8 mm

FIGURE 1. VVER-1000 MAXIMUM DENSITY STORAGE RACK MODULE DESIGN
(REGION 2 CONCEPT)

associates

Two different rack designs are utilized. One is a poisoned rack used for storage of fresh fuel, full core emergency discharge or fuel which has not achieved sufficient burnup (Region I). These channels are connected by a top grid and bottom base plate to include a space for water between the channels.

The second rack module (Region 2) uses the same poisoned channels but they are welded together at the corners to form a tight array, without water spaces. Fuel placed in Region 2 must have achieved a specified burnup depending upon its initial enrichment. Specific numerical values for this figure are fuel design and rack dependent. For a VVER reload batch of 4.5 w/o initial enrichment, a burnup of 30 GWD/T must be achieved before this fuel assembly can be stored in Region 2. The USNRC accepts normal administrative controls used in refueling operations to determine which fuel meets the burnup criteria. The burnup is based on normal core depletion calculations. It is not necessary to measure or physically confirm reactivity. Although not a requirement in the U.S., interlocks can be programmed into the fuel handling machine in order to permit administrative verification of specific assembly burnup.

Typically, Region 1 is sized to receive one core equivalent of fuel assemblies. The balance of the pool is racked with Region 2 modules spaced as tightly together and as close to the pool walls as possible. A typical pool rack layout for a VVER-1000 reactor provides about 820 Region 2 spaces in addition to the full core discharge Region 1 (163 spaces). This layout will accommodate about 15-20 years of reactor discharges for this type reactor.

To optimize space, minimize handling and provide maximum economy, rack module arrays should be 9x7 to 12x9. Smaller sets can be used to fill odd areas or to avoid pool impediments. Other factors that need to be considered in the rack module design and configuration are:

- Pool weight limits (overall)
- Pool floor and wall obstructions
- Location and number of leak channels
- Amount of fuel in pool during re-racking (removal strategy)
- Crane lift capacity
- Various criticality and thermal factors

For some pools, dose rate minimization at the outside of the pool walls (where personnel access is required) may be a consideration. The closeness of the stored fuel to the wall and the consequent reduction in water shielding thickness may result in higher dose rates at the outside of the wall. For the same reasons, gamma ray heating of concrete pool walls may cause local overheating of the concrete, especially when the full core has been discharged and pool temperatures are the highest. These problems can generally be avoided by storing only older fuel (with long-decayed fission products) in rack rows adjacent to the walls.

3.1 Poison Materials

As noted above, maximum density rack designs require the use of poison materials which is one of the historic problems in older US fuel racks. Three types of poison materials have been used in the US; Boraflex, Boral and Borated Stainless Steel (B-SS).

Boraflex is less expensive than Boral but it is more difficult to work with during manufacture. Neither material has performed well in the past, but Boral problems appear to be resolved. Boraflex, however, continues to deteriorate and presents a problem for utilities currently using this poison material. All new racks supplied in the US over the last 10 years use Boral inserts. Borated stainless steel, except for one early application has not been used in commercial reactors. It has been used in the navy nuclear program which is not subject to USNRC regulations and licensing procedures.

4. LICENSING REQUIREMENTS

In the US, rack designs must satisfy all requirements of primary safety related equipment for the nuclear plant contained in 10CFR50 App. A, General Design Criteria 61. The requirements for spent fuel storage facilities are delineated more fully in:

NRC Reg. Guide 1.13:"Spent Fuel Storage Facility Design Basis" and

ANSI/ANS 57.2-1983: "Design Requirements for Spent Fuel Storage Facilities at Nuclear Power Plants."

The regulatory process and procedures are specified in:

NUREG 0800:" USNRC Standard Review Plan, Section 9.1.1 - Spent Fuel Storage"

In general, the NRC requires detailed analyses in three areas:

Criticality
Thermal-Hydraulics and
Structural/Seismic

A criticality evaluation must show a neutron multiplication factor of less than 0.95 for all conditions of fuel storage and during installation of the racks. A thermal hydraulic study must show that the pool is properly cooled during the most heat inducing conditions possible during a full core off-load. A seismic evaluation describes dynamic loads induced by geological activity at or near the plant site. A structural analysis must show fuel rack strength capability to withstand combined static and dynamic loads and accidental drop loads.

Accident analysis requirements are prescribed by NUREG-0800 Standard Review Plans (SRP's), and these include:

Dropped fuel: At least two different dropped fuel scenarios are applicable to the racks that require a combined structural and criticality evaluation.

Misplaced fuel: If it is possible to place a fuel assembly outside the rack cell or in the wrong cell location, a criticality analysis must show that the multiplication factor is below the 0.95 level. A physical blocking device added to the rack can usually preclude the possibility of fuel misplacement.

Loss of pumped cooling service: If the spent fuel cooling system is a single or dual pump system, the loss of a single pump must be supported by analysis that will show that the fuel cladding temperature does not exceed the maximum temperature limit.

5.   FUEL RACK ANALYSIS AND METHODS

As stated above, the safety requirements have been set for many years and are well established. The NRC's interpretation of how one satisfies these requirements has evolved significantly over the years. Today very sophisticated analyses are required to prove conformance with requirements.

In the early days of fuel rack design, mainframe computer analyses were used. These were costly to perform and many simplifying but conservative assumptions were made to save computer and engineering costs. Also removal of conservatism -- particularly at high cost -- was not as important then as it is today. Now, fast running, personal work stations are available to perform sophisticated, parametric analyses quickly (on-line in many cases) and at low cost. These modern devices and techniques are used to remove previously imposed (conservative) assumptions and uncertainties. Typically, these modern calculations show considerable improvement in design margins relative to the established requirements; and this gain in margin can be used to upgrade the performance of the spent fuel pool and rack and, in turn, increase the in-plant storage capacity.

Most of the recent spacing gains have been made by advanced analysis with uncertainties and conservative assumptions removed. Explicit modeling of actual spent fuel pool phenomena has shown previous design calculations to be conservative. This newly gained margin is used to accommodate more fuel.

In particular, the use of hydrodynamic coupling analyses generally show that fuel racks in close proximity to walls or adjacent fuel racks will not impact each other with the high loads found to be evident in a dry analysis. In most cases, racks of similar size and weight will move together in phase. A successful analysis application will, thereby, reduce the need for large gaps between fuel racks and between fuel racks and walls to prevent impact by sliding or rocking rack motions. Obviously, with less gap required more fuel storage cells can be placed in the pool. In order to assure sufficient conservatism, the seismic analysis is performed with site specific seismic spectra or time history, with multi-rack computer modeling and with a variety of rack fuel loadings. A three dimensional, multi-rack, whole-pool analysis is preferred by the USNRC. Ground level accelerations are typically .25 g

6. REMOVAL AND INSTALLATION

The specific plan for rack removal and installation depends on the old rack structure and its attachments to the pool floor and walls. Installation of racks in clean, uncontaminated pools is cheap and fast. However, installation of new fuel racks and the removal of old racks become more difficult as the pool contains spent fuel. Below 50% the problem is not too severe.

The problems increase when the pool is more than 66% full. In this case, the first set of maximum density modules replace the empty old racks. Fuel is moved to the new racks, to the extent possible, and additional maximum density racks are installed. This procedure continues until all racks have been replaced. After the first new modules are installed, several empty rows are maintained to provide sufficient full core reserve as the process of removal and installation continues. A significant amount of fuel shuffling is required during the rack replacement in a nearly full pool. This situation should be avoided if possible, but with good planning and execution, the rack exchange can be accomplished without problem.

7. RE-RACKING SCHEDULE

The typical schedule for a US re-racking project is shown in Table 1. Design, analysis and licensing of a first time project will take six to nine months in the US --- even with a regulatory agency that has faced this issue many times and with suppliers which are well aware of the issues and how to address them. With an inexperienced designer and a competent authority which is considering these issues for the first time, the schedule could be double.

TABLE 1

RE-RACKING SCHEDULE

DESIGN, ANALYSIS & LICENSING: 6 - 9 MONTHS (1ST TIME)

RACK MANUFACTURE (1):   6 - 8 MONTHS

REMOVAL & INSTALLATION (2):   3 - 5 (--8) MONTHS

DECON., PKG., DISPOSAL (3):   1 - 3 MONTHS

---

(1) Max. density poison racks; 10 -12 modules, 1000 cells.
(2) Old pool has seismic restraints and is about 50 % full;
    One 10-hour shift per day.
(3) In drums for LLW disposal.
(*) Some activities can occur in parallel or overlap.

On the other hand, if the plants and spent fuel pools are standard -- or nearly so -- then generic approval can be pursued with subsequent pool licenses considering only site specific issues such as seismic response.

A medium sized, experienced manufacturer should be able to produce about 10 finished storage cells per day. Therefore a 1000 unit rack set will take about 100 working days or about 6 calendar months. Thus, rack sets for two reactors could be produced per year by such a manufacturer. More could be done by tooling up and committing floor space to the effort, but if a large number of reactors, say 8-10 units per year, are to be re-racked then more than one manufacturer will probably be needed.

Normally, the rack manufacturing can be started near the end of the licensing phase and end somewhat after the installation begins. For a very tight installation schedule, however, it would be best to have all modules on-site prior to start of installation.

The installation and removal schedule depends on many factors. The most important factors are:

- degree of restraint in the old racks

- amount of spent fuel in pool at start

  $<50\%$ - easy
  50-65%   - difficult
  $>65\%$ - very difficult

- degree of contamination in pool and old racks.

But, the most significant factor in US re-racking is planning and good pre-knowledge of the pool situation. The experience in the US with re-racking is that it will take about 3-5 months for a well planned campaign and as much as eight months for a poorly planned or problematic effort.

Decontamination, packaging and disposal will take one to three months and generally can be performed off the critical path.

8. RE-RACKING COSTS

Re-racking cost is a function of the size and complexity of the campaign. An example is given in Table 2.

TABLE 2

RE-RACKING COST ESTIMATE (1996 USD)
(Excluding internal utility costs)

DESIGN, ANALYSIS & LICENSING:	850,000
RACK MANUFACTURE :	2,000,000
REMOVAL & INSTALLATION :	1,350,000
TOTAL	4,200,000

Here we assumed that a rectangular pool with dimensions 30 x 40 feet is re-racked with 980 maximum density storage cells in 10 modules. We assume further that the old rack has seismic restraints and is 50% full at the start of the installation.

The estimates do <u>not</u> include internal utility costs for fuel handling, health physics, etc. They include the cost of packaging <u>for</u> disposal but they do <u>not</u> include the cost of disposal or disposal facilities.

Training and transfer of technology can be made during the first campaign so that future sets can be made in the Region. We estimate that these can be made in the Region at about one-half to one-third the first time cost. Design cost for identical sets at the same site would be minimal for future sets.

The cost of storage in maximum density storage racks then is:

> First-time: $4.2 million / 980 cells = $4,200/cell
> Subsequent: $2.5 million / 980 cells = $2,500/cell

This compares favorably with dry storage casks (eg $.35 million/ 20 cells* = $17,500/cell), for example.

In addition, re-racking of all pools would postpone the need for dry storage by 10-15 years and would make future dry storage facilities cheaper because the fuel to be stored at that time would be cooled for ten or more years.

9.   SUMMARY AND RECOMMENDATIONS

Modern maximum density fuel rack designs have tightly spaced arrays which approximate the reactor core pitch ($\sim$ 240 mm pitch for VVER-1000). The basic cell is comprised of eleven gage 316-SS with boral poison inserts for reactivity control. The rack modules have minimal space between them and are placed as near to the pool wall as possible. Every available space is utilized and most pools can achieve sufficient storage capacity for 15-20 years of reactor operation (life-of-reactor in some cases).

The total cost of a re-racking campaign in the Region should be less than $5,000 per storage space and well under the next cheapest alternative.

For utilities contemplating a re-racking campaign we recommend the following:

(1) Go directly to maximum density racks. (The major cost is in the design, licensing, installation, removal and disposal. These items cost about the same regardless of the number of cells so why not get the maximum capacity for the expenditure.)

(2) Standardize the design as much as possible for overall ease of licensing. Allow at least one year for the first approval.

(3) Use proven materials.

(4) Start the re-racking campaign before the pool is 50% full, if possible.

(5) Prepare very detailed plans for the removal and installation program.

(6) Use experienced installation/removal teams.

(7) Allow 18-24 months for the first campaign; 12-15 months for subsequent efforts.

(BNFL) practices on wet fuel storage at Sellafield

NDS
Nuclear Fuels plc, Sellafield, UK

SYNOPSIS

A wide range of irradiated nuclear fuels are stored in ponds at BNFL, Sellafield, prior to reprocessing. Magnox fuel from the first UK power generating reactors has been stored and reprocessed since the mid 1950's and the reprocessing of AGR and Water Reactor fuels has recently commenced in Thorp (Thermal Oxide Reprocessing Plant). The paper summarises the irradiated fuel management strategies that have been adopted for each type of fuel with the aim to minimise or prevent corrosion during pond storage. Current development work in support of the long term storage of AGR fuel is also outlined.

1. INTRODUCTION

A wide range of fuels are stored at Sellafield in water filled pools prior to reprocessing. The pond water acts as an infinitely flexible transparent shield which facilitates remote handling to be carried out from above the pond surface. Early fuel was stored in open skips in uncovered ponds, which resulted in chloride contamination of the pond water by chloride ions entrained in the air of the coastal environment. Under these conditions the cladding materials were susceptible to corrosion. However, pond storage was short term and the fuel was generally reprocessed before penetration of the cladding occurred.

As it was recognised that any fuel corrosion during periods when reprocessing was not possible would cause operational problems it was decided to develop an ability for the long term storage of fuel. This meant that the water chemistry within a pond needed to be improved to prevent corrosion of the cladding. Development of reliable long term storage regimes has taken place over the past sixteen years. BNFL has concentrated on the concept of containerised storage, now used for all types of fuel stored at Sellafield, and the optimisation of the pond water chemistry.

The first tranche of storage operations involved Magnox fuel (natural uranium clad in magnesium alloy) from the eleven British Magnox stations: augmented by Magnox fuel from the Tokai-Mura reactor (Japan) and the Latina reactor (Italy), a total of twenty eight reactors.

Irradiated nuclear fuel from the second generation of British commercial reactors - Advanced Gas Cooled Reactors (AGRs) - is also stored at Sellafield and is now being reprocessed in Thermal Oxide Reprocessing Plant (Thorp). This fuel comprises uranium dioxide pellets clad in stainless steel.

Also stored at Sellafield is irradiated fuel from water-cooled and water moderated reactors (mostly non-UK) which is also now being reprocessed in Thorp. This fuel is typically uranium dioxide pellets clad in zirconium alloy. Whatever the source of the fuel, the storage regime should satisfy the following criteria:-

i   Fuel clad integrity should be maintained to eliminate, as far as is practicably achievable, radioisotope release to the environment.
ii   The fuel must be shielded to minimise radiation up-take by plant operators.
iii The fission product thermal output of the fuel should be dissipated to prevent excessive fuel temperatures during storage.
iv   Storage must be such as to prevent a critical assembly being formed under any condition.
v   Radiation stability of the storage environment should be such that breakdown products can be easily controlled.

Based on the considerations of (i) to (v) above the decision was taken at Sellafield to store early Magnox fuel in deep water filled ponds. Subsequently, this storage philosophy was adopted by the British electricity generating utilities for the majority of their commercial Magnox fuel stations. More recently, water storage has been chosen for irradiated oxide fuels in Britain, and by most other countries which have a significant nuclear contribution to their power generation.

2. FUEL STORAGE PONDS AT SELLAFIELD

There are four main fuel storage ponds at Sellafield. Fuel storage at Sellafield is shown schematically in Figure 1.

Fuel Storage at Sellafield

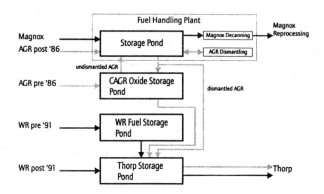

Figure 1. Fuel storage at Sellafield.

These ponds are unlined (except for a stainless steel clad wind-water line), reinforced concrete structures built with an 'above ground' philosophy. Some pond walls are painted concrete. BNFL do not use metal clad ponds, but this topic has been carefully researched and the modern concretes employed have been shown to have negligible corrosive ion leaching and permeability to water.

Sellafield, historically, used a policy of once-through pond water purging and subsequent discharge to sea because of the abundant supply of excellent quality water in the English Lake District and Sellafield's coastal location. This situation has now been improved to the position where incoming water is de-ionised and filtered and most outgoing water is passed through a Site Ion Exchange Plant (SIXEP) before sea discharge to minimise release of radioactive species to the environment. Economies in water usage are achieved by purging from pond to pond where possible.

The fuel storage ponds are purged by removing the bulk pond water at a rate of approximately 6% of the total pond volume per day. A constant pond water volume is maintained by adding inactive demineralised water to which chemical dosing may have been performed depending on the type of fuel being stored.

Containerisation can be an effective method of reducing the escape of radionuclides into the bulk pond water during storage. Figure 2 shows a container for storing Magnox fuel. AGR containers are almost identical in design.

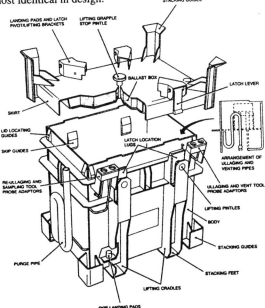

Figure 2. Magnox storage container.

Before containers are opened they can be flushed to remove the accumulated activity and the resultant effluent treated prior to discharge. However, some activity can be released into the water during routine pond operations. Thus the pond purge water contains radionuclides as both suspended solids and dissolved salts. The suspended solids are removed from the purge by sand filtration and the isotopes of caesium (a significant soluble component) and strontium are removed by ion exchange prior to sea discharge of the purge. These processes are performed in SIXEP.

Depending on the type of fuel being stored the pond water may be dosed with inhibitors to reduce fuel corrosion. The various water chemistries used at Sellafield are summarised in Table 1.

Pond	Magnox Fuel	AGR Fuel	WR Fuel
Fuel Handling Plant	Ullaged containers pH 13	Flooded containers pH 11.4	
AGR Oxide Storage Pond		Flooded containers pH 11.4	
WR Fuel Storage Pond			Ullaged/vented MEB's pH 7
Thorp Storage Pond		Ullaged containers pH 7	Ullaged/vented MEB's pH 7

Table 1. Fuel storage conditions at Sellafield.

3. MAGNOX FUEL STORAGE

Sellafield currently accepts irradiated fuel from 13 Magnox power stations in the UK and elsewhere, and Magnox fuel will continue to arrive at Sellafield beyond the year 2000. Temporary dry storage has been successful at the largest British Magnox power station, Wylfa, but other stations are not expected to convert from pond storage. Dry storage at Sellafield following wet storage at reactor has been assessed and is not considered to be practicable.

Irradiated Magnox fuel is stored at the power stations in open topped skips and transported to Sellafield in shielded flasks. At Sellafield, in the Fuel Handling Plant (FHP), the fuel is placed into containers. Because of the radiochemical conditions caused by the presence of Magnox it has proved possible to have these containers ullaged. With a sealed container it becomes feasible to engineer a water chemistry within the container that is optimised at reducing the corrosion of the Magnox cladding. Based upon an extensive research programme a water chemistry of pH 13 and $(Cl^- + SO_4)<0.5$ ppm was developed for these containers. Additional control of the Magnox corrosion rate is obtained by cooling the pond water to 15-17 °C. Under these conditions storage times of at least five years without cladding penetration are expected to be achievable.

The containers act as a barrier to any radioactivity released from failed fuel elements which can occur in the reactors or from delugging and desplittering operations. Any corrosion debris produced during storage are also contained and can be removed via a skip desludging machine.

4. ADVANCED GAS COOLED REACTOR (AGR) FUEL

Upon discharge from the reactor AGR fuel elements are stored in open skips in ponds at the reactor stations which are dosed with boron for criticality control. After a minimum of 90 days cooling these skips of fuel are transported in a flask to Sellafield where they are placed inside lidded containers using a dry inlet facility shared with Magnox fuel in the FHP. The container lid provides criticality control by segregation so no boron additions are necessary to the pond water at Sellafield. The design of the lid allows containers to be triple stacked. After a minimum of 180 days cooling the elements are dismantled, the fuel pins being transferred into slotted cans and the redundant graphite sleeves and additional stainless steel components being stored in drums as Intermediate Level Waste (ILW). This not only converts the fuel into a form suitable for reprocessing but also results in a 3 or 4 fold increase in storage density. After dismantling the fuel is transferred either to an intermediate pond for further interim storage and thence to Thorp or the fuel can be transferred directly to Thorp.

Because of the recognised risk of fuel pin cladding corrosion due to thermal and radiation induced sensitisation, the condition of AGR fuel elements has been subjected to an extensive monitoring programme to determine the effects of increasing storage time (the longest stored fuel is approximately 18 years) and increasing fuel burn-up (initial fuel was approximately 5000 MWd/t stringer mean irradiation (SMI), now at 27,000 MWd/t SMI, with a planned increase to 30,000 MWd/t SMI and more). Fuel elements were removed from the pond and subjected to detailed destructive examination. From this work a thorough understanding of the pond storage behaviour of AGR fuel has been obtained. This programme also resulted in the identification of a corrosion inhibitor. It was found that by dosing the water to pH 11.4 using sodium hydroxide, the failure of AGR fuel pins, which otherwise would have perforated due to the pond water chloride levels (2-4 ppm), could be prevented. All the main AGR fuel storage ponds now use dosed pond water.

The design of the container and lid are such that a gas space (ullage) can be formed in the lid thus isolating the container contents from the bulk pond water. However, it has been found that radiolysis of water caused by storing short cooled fuel in dosed water can potentially result in the formation of atmospheres of hydrogen and oxygen gases in the ullage. Most AGR fuel is therefore currently stored with no ullage in a condition that allows free interchange of water between the container and bulk pond water which is dosed with hydroxide. The exception to this is AGR storage in the Thorp storage pond. Here it was recognised that, due to considerations of compatibility with Boral in the Multi Element Bottles (MEB's) used for storing Water Reactor (WR) fuel, sodium hydroxide dosing could not be used. In this pond ullaged containerised storage has been successfully achieved by the development of catalytic recombiners in the ullage space which prevent the formation of explosive atmospheres. The containers are filled with high quality demineralised water (~0.1 ppm Cl). It has been demonstrated that at these very low chloride levels the lack of hydroxide dosing does not result in fuel pin corrosion. Prior to the operation of the Thorp storage pond an ullaged container holding undismantled fuel was established with low chloride undosed water and liquor sampling demonstrated that fuel pin perforation had not occurred after 700 days storage. When the Thorp storage pond became operational the first dismantled AGR fuel to be stored was also monitored and there was no indication of fuel pin perforation after ~1100 days storage.

5. WATER REACTOR FUEL STORAGE

On discharge from the reactors the fuel is stored for an initial cooling period in a pond at the reactor site. BWR ponds are filled with demineralised water, whereas PWR ponds are filled with a dilute (approximately 0.2M) boric acid solution. The boric acid in PWR pools results from the mixing between the boronated water in the reactor vessel and the bulk pond water during refuelling operations. This chemistry results in BWR ponds operating at approximately pH 5.8-7 and the PWR ponds at approximately pH 4.5 -6. Most of these ponds operate at, or below, 40°C.

The fuel is transported from the reactor ponds to Sellafield in MEB's contained within heavily shielded, high integrity, transport flasks. The MEB's are cylindrical stainless steel vessels containing stainless steel clad Boral dividers between the fuel assemblies to prevent criticality. Boral consists of boron carbide particles in an aluminium matrix clad with pure aluminium and is widely used as a neutron absorber. A typical MEB for storing BWR fuel elements is shown in Figure 3. MEB's are primarily used to minimise contamination of the transport flask from crud or spalling surface layers.

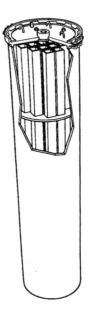

Figure 3. Multi Element Bottle.

On arrival at Sellafield, the flask is placed in the pond, opened under water, and the MEB containing the fuel removed. The flask is then removed from the pond, decontaminated, and returned to service, while the MEB is transferred to a storage frame which supports it vertically during its storage period in the pond. The fuel storage ponds used for the storage of WR fuel contain undosed demineralised water with a purge to maintain low chloride plus sulphate concentrations (<0.5 ppm).

Several benefits accrue by the use of MEB's; these include easier fuel handling with less risk of damage to the assemblies and less contamination of the storage pond. They also allow control of the water chemistry around assemblies. Boral inserts allow the close packing of assemblies within the MEB. In all current designs of MEB's there is exposed Boral which results in the removal of oxygen, produced by radiolysis of the water in the MEB, to produce an ullage gas composed mainly of hydrogen and nitrogen. However, in a few early designs the Boral is wholly clad in stainless steel and this does not remove oxygen, resulting in an oxygen/hydrogen mixture in the ullage. Consequently these MEB's are vented to the pond water. Thus MEB's can contain either reactor pond or Sellafield pond water.

The newest designs of MEB's use boron loaded stainless steel for criticality control and this has allowed the quantity of fuel stored to be increased by 20%.

The safety case for the storage of Mixed Oxide Fuel (MOX) has recently been completed. The ability to store MOX fuel completes and closes the MOX fuel cycle - manufacture, irradiation, storage, reprocessing and recycle of uranium and plutonium.

No evidence of pond-induced degradation of WR fuel has been found at Sellafield; this reflects the world experience in the storage of water reactor fuel.

6. RADIOLOGICAL CONDITIONS

BNFL has a continuing commitment to reduce radioactive discharges from all sources at Sellafield. The significant reductions associated with pond storage operations have been achieved by the combination of the introduction of SIXEP and the improvements in pond water chemistry and containerisation of the fuel.

The success of a fuel storage regime can be judged by the amount of radioactivity discharged in the pond purge and, in general, as each new pond has been commissioned an order of magnitude reduction in pond water activity levels have been achieved.

This is best illustrated by consideration of ^{137}Cs levels in the pond water this being the most significant soluble fission product. The FHP routinely operates with a caesium pond water activity of 300-400 Bq/ml with this activity arising almost entirely from the handling of Magnox fuel with mechanical damage. Prior to the adoption of containerised storage pond water activity levels were of the order 1000-3000 Bq/ml.

In the intermediate buffer storage pond for dismantled AGR fuel pond water activity levels are typically in the range 30-40 Bq/ml. This reduced activity level reflects the much lower activity release rates from oxide fuels and arises as a result of the perforation of a small proportion of the AGR fuel stocks before pond water dosing was introduced and the storage of this fuel in vented containers. In contrast activity levels in Thorp storage pond are of the order 4-5 Bq/ml as a result of the use of sealed containerised storage for both WR and AGR fuel.

7. RECENT / ONGOING RESEARCH AND DEVELOPMENT IN SUPPORT OF POND STORAGE

7.1 Magnox Fuel

No work in support of pond storage is currently being carried out. The current storage regime for Magnox fuel is considered to be more than adequate and further development work is not required.

7.2 AGR Fuel

The use of catalytic recombiners has allowed, for the first time, ullaged storage of AGR fuel. A programme consisting of the monitoring of 20 containers in THORP Receipt and Storage pond demonstrated that the catalysts prevented the formation of explosive atmospheres in the ullage space. This programme is continuing to extend the operational data to fuel with a wider range of burn-ups/cooling times and to quantify achievable catalyst lifetimes.

A mathematical model has been developed that predicts activity release from AGR fuel stored in one of the ponds at Sellafield. By comparing predicted with actual activity release it is possible to show that pin perforation is not occurring. This is an extremely cost effective means of monitoring a large quantity of fuel.

Experimental work has demonstrated that short term deviations from optimum pond water chemistry i.e. increased chloride levels, do not result in fuel corrosion when dosed ponds are used. A container of AGR fuel was isolated from the pond by ullaging and the chloride level inside the container progressively increased. Fuel integrity was monitored by taking liquor samples and measuring activity levels. The were no indications of fuel failure even after ~100 days at 30 ppm chloride.

Currently Thorp storage pond can accept AGR fuel with an irradiation of 18,000 MWd/t and 5 years cooling. Work is in progress with the aim to be able to receive fuel up to 31,000 MWd/t and 6 months cooling.

Additional experimental work is to be performed to increase confidence in the ability to store AGR fuel for extended periods without pin perforation. The average storage time of AGR fuel currently stored at Sellafield is 6½ years and the maximum storage time in dosed pond water is 10 years. The aim of the proposed work is to validate storage regimes for much longer timescales and is in response to BNFL having accepted contracts for the storage of Scottish Nuclear AGR fuel for up to 80 years.

7.3 WR Fuel

Work has demonstrated that accumulated activity inside a MEB can be removed by flushing. MEB's will be flushed prior to their opening in Thorp Head End to reduce activity release to the bulk pond water.

Very early receipts of WR fuel are being transferred from open pond storage into MEB's prior to transfer to the Thorp storage pond. Recently eight assemblies, with storage times of between 17-21 years have been visually inspected during transfer. Some of these assemblies were examined previously nine years ago and no change in the condition of this fuel was observed.

Activity levels associated with used MEB's are being monitored so a policy on future use, either re-use or disposal, can be formulated.

8. CONCLUSIONS

The sodium hydroxide dosing of the water surrounding both Magnox and AGR during pond storage acts as an effective corrosion inhibitor. Storage in very low chloride undosed water has also been shown not to promote the corrosion of AGR fuel. Containerisation provides an engineered means of controlling both the initial and long term water chemistry for fuel storage.

Containerisation of fuel provides an effective barrier between the fuel and the bulk pond water. The diffusion of both soluble and solid activity is controlled in the case of ullaged storage. This reduces the amount of radioactivity in the bulk pond water and consequently reduces the radiation dose to the people who operate the pond systems.

Treatment of liquors from pond systems in the SIXEP plant has a major effect on reducing the amount of radioactivity discharged to the environment.

BNFL has built on the extensive experience and skills of its personnel to provide modern irradiated fuel storage and handling facilities which enable high levels of throughput whilst minimising fuel corrosion, radiation doses and environmental impact.

uccessful progress in the Post Operational Clean Out of ne Magnox ponds and silos at Sellafield

V MILLER BSc and E J WILLIAMSON BA, MRSC, CChem, MBNES
tish Nuclear Fuels plc, Sellafield, UK

ABSTRACT

British Nuclear Fuels Plc (BNFL) owns and operates the UK's Nuclear Reprocessing Facility at Sellafield in West Cumbria. These nuclear reprocessing operations have been carried out over the last 40 years and three generations of plant have been successfully employed.

Prior to decommissioning plants which have reached the end of their operational lives are shut down safely and Post Operational Clean Out (POCO) is carried out to reduce the radioactive inventory to a point where large scale Decommissioning, Decontamination and Dismantling can begin.

This paper describes the successful implementation of POCO in three areas within the historic Magnox Ponds and Silos. These are:

- Pond Fuel Recovery
- Fuel Cladding Swarf Retrieval
- Sludge Retrieval

INTRODUCTION

BNFL owns and operates the UK's Nuclear Reprocessing Facility at Sellafield in West Cumbria providing the world with the best nuclear fuel services available. These nuclear reprocessing operations have been carried out successfully over the last 40 years and three generations of plant have been used in this time.

Prior to Decommissioning plants which have reached the end of their operational lives are shut down safely. A care and maintenance period follows where Plant Improvement Projects are carried out to maintain or improve standards (eg. ventilation systems, services etc). This is necessary to facilitate the next phase of operations, Post Operational Clean Out (POCO) of the plants. The purpose of POCO is to reduce the radioactive inventory to a point where large scale Decommissioning Decontamination and Dismantling can begin.

One group of plants undergoing POCO are the historic Magnox Ponds and Silos at Sellafield. These plants and projects are operated by the Retrievals Department which is part of the Waste Retrievals and Decommissioning (WR&D) division of the BNFL UK Group. During routine operations over the lifetime of these plants, over 25 000 tons of metal clad natural Uranium fuel from UK, Overseas customers and BNFL has been processed. This processing gave rise to both liquid and solid wastes of various types and a high proportion of the solid waste is Intermediate Level Waste (ILW). ILW must be retrieved for conditioning then interim storage and eventual export to deep disposal. All of this will be carried out safely, efficiently and with care for the environment whilst at the same time delivering cost effective solutions for our customers.

BNFL has developed an overall Site ILW strategy which has been planned and agreed with the Regulators and our customers. It is however a strategy which is being continuously improved as part of the BNFL Strategic Management Process (SMP). WR&D has an ongoing programme of work to develop and to implement techniques for the retrieval of these different types of wastes. This paper describes some of the work that is being successfully carried out in these challenging areas.

POND FUEL RECOVERY

Background

The historic Magnox fuel receipt, storage and decanning buildings have handled in excess of 25 000 tons of irradiated metal clad natural Uranium fuel. The fuel that has been handled has come from a variety of customers, from both in the UK and overseas. These facilities ceased main stream operations in 1986 and there remained in the storage ponds approximately 1 000 tons of fuel stored in containers called skips most of which were not easily accessible. Much of the fuel was also was in a corroded condition. In addition there had been the accumulation of Magnox sludges, from corrosion of the fuel cladding, in areas of the storage pond. This must be removed to allow more POCO work to be carried out in the Pond.

This fuel must be recovered to enable its reprocessing and to reduce the overall radioactive inventory of the plant. This will help to reduce the radiation dose to the workforce.

The Solutions

The installation in 1991 of a new specialized overhead crane system, the Pond Skip Handler, to replace an existing one. The design of which was improved so that it could accept a number of interchangeable mast components to assist with the different retrieval operations required in the storage Ponds.

In common with other projects extensive design studies were undertaken to ensure that the installation of the equipment could be carried out in the safest way and with the radiation doses to the workforce being consistent with the As Low As Reasonably Practicable (ALARP) principle. The initial installation of and commissioning of the skip handler was completed, there have been continuing improvements in its associated equipment which have led to the enhancement of its overall capability.

These enhancements have taken the form of :

A shielded capsule which allows operators to work very close to the Pond surface

A skip recovery grab which allows recovery of mis-aligned skips

A sludge grab to recover bulk sludges from the Pond

A fuel grab to recover loose fuel from the bottom of the Pond

The skip handler is shown in operation in figure 1.

Figure 1 Pond Skip Handler

Achievements to Date

The operation of the Skip Handler is routinely fully remote and is controlled from a remote control cubicle by a single operator using a PLC and video system. The operation of the skip handler in this way has considerably reduced the radiation dose to the workforce. This reduction is of the order of 250 man mSv per year which is equivalent to over 16 man years dose based upon the current internal Company dose limits which prevents the workforce exceeding 75mSv in a 5 year period.

The use of the shielded capsule and the skip recovery grab have enabled the plant to recover over 200 mis-aligned skips so allowing their inspection and the recovery of fuel residues from them for reprocessing.

The recovery of bulk sludges from the Pond is continuing with both skip washing in the Rotary Skip Wash (RSW) and sludge grabbing contributing to a total since shutdown of the plant in excess of 400 M^3 .

Fuel recovery has progressed with approximately 750 tons of fuel removed, most of the reprocessible fuel that remains in the Pond will require further treatment before it can be sent to this route.

Both the recovery of fuel and bulk sludges from the Pond has helped to reduce the overall radioactive inventory of the plant as shown below in figure 2. This will also help to reduce the overall radiation dose in the environment of the plant.

Figure 2 Reduction in Stock Expressed as Total Activity

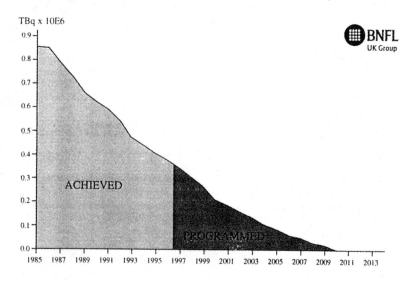

FUEL CLADDING SWARF RETRIEVAL

Background

The Magnox fuel metal cladding is removed (decanned) before the Uranium metal is sent for reprocessing. In this decanning process the cladding is removed as a metallic (magnesium alloy) swarf. This material is ILW in nature due to activation products and Uranium attached to it. The swarf storage silos came into operation in 1964 and were progressively extended until the late 1980's to a total of 22 silos or compartments. Figure 3 shows the building containing these silos.

Figure 3 Isometric View of the Building Containing the Wet Storage Silos

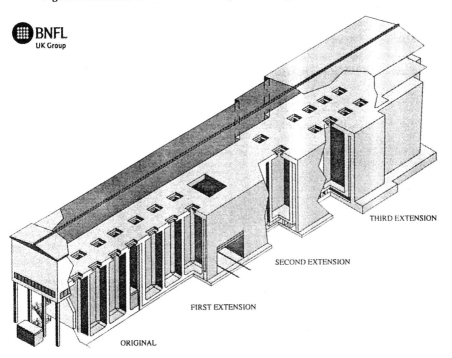

As part of the ILW strategy it has been BNFL's aim to transfer all the Magnox swarf that can be directly encapsulated to the Magnox Encapsulation Plant (MEP). This plant commenced operations in 1990 and so new arisings of swarf have been sent there for encapsulation in a cement grout in 500 litre stainless steel drums. These drums are stored in a purpose built interim store prior to their final deep disposal.

The retrieval of swarf from the silos has concentrated on the ones in the more modern extension of the plant, which comprises silos 19 to 22, where the swarf is suitable for direct encapsulation because of the relative lack of associated sludge or mixed ILW.

The Solution

The Swarf Retrieval Facility (SRF) which was designed for silos 19 to 22 is shown below in figure 4.

Figure 4 Swarf Retrieval Facility

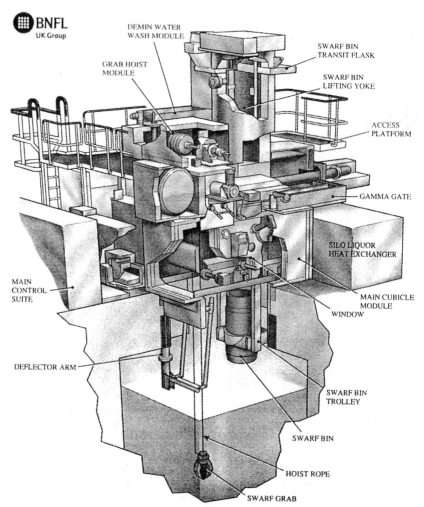

The equipment was designed by a combined integrated project team of design and operational personnel to ensure that the safety and operability issues adhered to the ALARP principle. A number of important features were designed into the equipment these were :

- Designed to be minimum maintenance and in a modular form because of the potentially high radiation and contamination. Any failed equipment can be removed easily.

© IMechE 1996 C512/

- Independent seals providing an interface between the equipment and the silo ventilation system. This to prevent the spread of contamination and the potential buildup of Hydrogen.
- Weight optimization to minimise loading on the silo roof loading whilst meeting modern shielding requirements.
- Equipment can be retracted into the facility inside its lower shielding so that when it is was moved between silos there was a minimisation of dose uptake.
- A unique lifting beam to allow the movement of shielded transport flasks on to and off the SRF.

Achievements to Date

Due to the integrated project approach carried out involving the fabrication and testing off Site enabled installation procedures to be developed, test procedures and modifications to the facility to be carried out in an inactive environment. In addition operations and maintenance personnel could be trained to work on the SRF. All of these measures benefited the project as radiation doses for its installation and subsequent maintenance have been much lower than predicted.

The SRF was delivered, installed and commissioned ahead of schedule. It is currently exceeding its targets for swarf retrieval, it has emptied one silo in less than 14 months, 1 month of this was a routine maintenance shutdown, the target for this work was expected to take 18 months.

SLUDGE RETRIEVAL

Background

As part of the reprocessing operations on the Sellafield Site irradiated nuclear fuel is received and stored in cooling Ponds underwater. Over the lifetime of the receipt and storage plant corrosion of the Magnox fuel can has led to the accumulation of quantities of radioactive Magnox sludge within the Ponds and its associated underwater bays. Some of this sludge is also to be found in the effluent settlement tanks in an adjacent plant where liquid effluent was treated prior to its discharge.

The requirement to retrieve these sludges is due to the fact that they give rise to significant radiation levels, which in turn limits the working time in some areas of the plants due to the potentially high radiation doses to the workforce.

The Solutions

Each of these sludge accumulations pose different problems and hence a range of projects have been initiated. Each project requires extensive Design studies to be carried out to meet the technical objectives of the project whilst maintaining radiation doses in the ALARP principle.

To achieve this, a series of inactive trials were conducted both at the R&D stage and also during off site construction where full scale simulations were used. These steps have brought significant advantages to the projects :

- Installation, operating and maintenance personnel could be trained in an inactive environment
- Significant amounts of initial testing could be carried out to review safety issues and to satisfy regulatory bodies
- The optimization of operating and maintenance procedures

Figure 5 illustrates the equipment used for the bulk recovery of sludge from the two effluent settling tanks where approximately 200 M^3 of sludge had accumulated. These tanks are adjacent to the Decanning facility and storage Ponds. Figure 7 illustrates the equipment that was installed to recover sludge from one of the wetbays in which approximately 230 M^3 of sludge had been pumped as an interim storage measure early in the life of the Decanning facility.

Figure 5 General View of Settling Tank Desludging Project

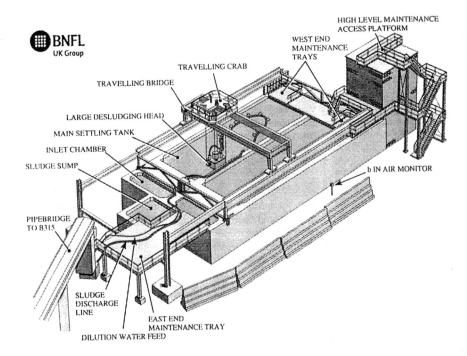

Figure 6 Effluent Settling Tank Desludging History

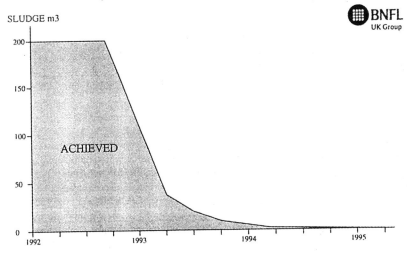

Figure 7 General View Showing a Section Through the Wet Bay Desludging Project

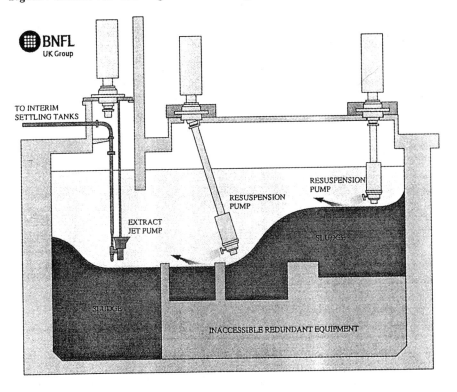

Whilst these projects are engineered differently to meet the requirements of the local environment they both work on similar principles. This is to hydraulically resuspend the Magnox sludge which converts the sludge to a slurry which can then be pumped via coaxial pipelines in a shielded pipebridge to intermediate settling tanks for measurement. It is then hydraulically resuspended again for its transfer to the Site Ion Exchange Plant (SIXEP) for interim storage in dedicated 1 000M³ tanks. The sludge will be eventually transferred as one of the feeds to the Sellafield Drypac Plant (SDP) for conditioning and then to the Waste Encapsulation Plant (WEP) for final encapsulation.

Achievements to Date

Installation of the settling tank desludging project began in April 1992 and was completed by mid August 1992. It then began operation and desludging output exceeded expected target figures. After 18 months and the removal of approximately 200 M³ sludge only minimal amounts remain. Development work is now in progress to remove the residual sludges and to enable the full Decommissioning of the facility.

The wet bay desludging project began installation in April 1995 and was commissioned in September 1995. Active commissioning to date has already successfully removed approximately 100 M³ sludge. Retrieval of the sludge will continue until only residual amounts remain and then decommissioning of previously installed and currently inaccessible decanning equipment in the wet bay can begin.

Both of these projects have been successful. Each of them, due to the Design, R&D, Safety and ALARP studies along with detailed work programming which were carried out prior to the installation work. Installation was completed in approximately half the anticipated radiation doses. In addition they have both exceeded their target retrieval rates and operations teams have been able to respond to maintenance and breakdowns promptly, due to the experience gained during installation, testing and commissioning.

Figure 8 Total Desludging Programme

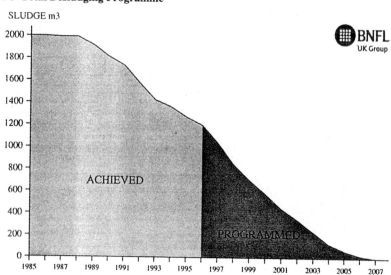

© IMechE 1996 C512/

THE FUTURE

The experience gained from these projects will be of great importance to the next series of projects that will be installed to continue the POCO and decommissioning of the Magnox Pods and Silos.

Within the decanning facility and associated storage ponds there is the need to continue to retrieve the sludge and fuel residues as well as clearing amounts of ILW stored underwater in the wetbay areas. The later work will require the installation of size reduction equipment and shielded transfer routes to the main storage pond. The work on the first wetbay installation is nearly complete with commissioning due later this year. This work will eventually allow the export of the ILW to the new ILW conditioning and encapsulation plant being built.

The challenge in the wet silo plant will be to install additional Silo Emptying Plants (SEP I, SEP II & SEP III) which will be used to retrieve the mixed swarf, sludge and other ILW from the silos. The designs for these plants are well advanced and the plants will be manufactured and installed by the year 2001.

In this paper we have also mentioned that two new major downstream facilities for the conditioning and encapsulation of the ILW are being currently designed and constructed, these are the Sellafield Drypac Plant (SDP) and the Big Box Treatment Plant (BBTP). These will be available in 2001 / 2002 which will enable them to receive and treat the output from the POCO and decommissioning activities. Once the ILW has been conditioned and encapsulated they will be stored in dedicated stores associated with these new plant prior to its final deep disposal.

CONCLUSIONS

The work outlined in this paper illustrates BNFL's Structured Systems engineering approach to the safe retrieval and treatment of ILW prior to its long term storage and final disposal. The success of this work is based upon the team approach which brings together designers, operators and maintainers throughout the life of a project. In addition the use of initial testing and commissioning in inactive environments has proved to be a very effective method in reducing radiation doses and gaining valuable experience for installation, operation and maintenance.

Overall the experience gained so far demonstrates that BNFL has the ability to undertake difficult and complex waste retrieval and management's projects in the UK and elsewhere in the world.

er-relationship between design and end-of life nditions of spent fuel and back end of fuel cycle uirements

EHS KTG, **V DANNERT** KTG, and **M WICKERT**
ENS AG Power Generation Group (KWU) – Nuclear Fuel Cycle, Germany

Synopsis:

Spent LWR fuel assemblies will have an increasing burn up. Target batch average burn-up of 55 GWd/tHM will be reached in the near future. Wet storage performance of spent fuel is scarcely impacted by increasing burn-up. Dry storage however needs a detailed assessment. Therefore the loading schemes of transport and storage casks need to be optimized based on individual fuel assembly end of life data. Individual fuel assembly encapsulation at the end of the wet storage period provides a number of advantages e.g. the decoupling of front end from back end requirements of the fuel cycle. Also, very late spent fuel conditioning for final disposal can be omitted.

1. INTRODUCTION

Spent fuel management has always been one of the most important stages in the nuclear fuel cycle. It is still a vital question to all countries with electricity producing power reactors. Spent fuel management begins with the discharge of spent fuel from the reactor and ends with final disposal of the spent fuel or of the residues from reprocessing of the spent fuel. Two main options for closing the fuel cycle exist at present: an open once-through cycle with direct disposal of the spent fuel and a closed cycle with reprocessing of the spent fuel and recycling of the Pu and U in new fuel assemblies. The "wait and see" option might be considered as a third for those who cannot decide for different reasons between reprocessing or direct disposal. The selection of a spent fuel cycle strategy is a complex procedure in which many factors have to be weighted: technical, political, economic and safeguard issues as well as protection of the environment.

To describe the scope and type of spent fuel management it has to be noted that worldwide 424 power reactors were operating at the beginning of 1993. The total amount of accumulated spent fuel was 125 000 thM. A great deal of that fuel has been reprocessed or foreseen to be reprocessed until now. Nevertheless, in the year 2000 the amount of accumulated spent fuel to be stored is expected to reach 150 000 thM. The global view shows a clear partitioning between reprocessing and direct disposal. The national view teaches that the individual nations adhere mostly either to reprocessing or to direct disposal [1].

Reduction of the fuel cycle costs by minimizing the amount of spent fuel discharged from the reactors by burnup increase, is an outstanding example for the cross links between the front end and the back end of the fuel cycle [2]. However, the advantage of decreasing amounts of spent fuel per unit of electrical power generated is accompanied by the possible drawback of an alteration of the end of life (EOL) conditions of the discharged fuel with increasing burnup. To a large extent wet fuel storage can accommodate the implications of increasing burnup. However, assessing dry storage of spent fuel indicates a possible influence from EOL-conditions of a high burnup spent fuel on dry storage performance.

2. THE BURNUP INCREASE OF SPENT LWR-FUEL

The increasing competition between different electricity generating systems is challenging the generating cost of nuclear power plants. Therefore, there is a strong incentive to improve the nuclear fuel cycle cost.

Further reductions on the front end of the fuel cycle cost and especially on fuel procurement are limited because of the competition between suppliers already experienced for decades. However, remarkable amounts are realizable by optimizing the design and the back-end expenses. This is valid for both for the once through cycle with Uranium fuel assemblies and for the reprocessing cycle with MOX fuel assemblies [3].

In particular, increase of discharge burnup gives LWR fuel a contribution which is significant. The increase of discharge burnup, now in the range of between 40 and 50 GWd/tU to 60 GWd/tU, will reduce the fuel cycle cost by about 10 % - 20 %. The technical development of cladding material and fuel rod design allows such an increase of burnup.

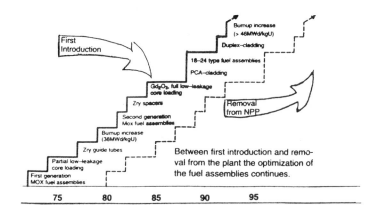

Fig. 1 Development of SIEMENS PWR fuel assemblies (FA)

Figure 1 shows the development of Siemens PWR fuel assemblies. The development comprises a plurality of improvements for getting better fuel performance, better fuel usage and last but not least increased burnup. In the early eighties the burnup reached about 40 GWd/tHM. Until the end of the nineties batch average burnup of 55 GWd/tHM will be generally achieved for both PWR and BWR-FA.

Figure 1 also describes the relationship between market introduction of improved FA and the time to receive those FA in the back end of the fuel cycle. It takes 7-10 years from loading a FA with improved design features until removal of such FA from the reactor. The time period to deal with the design changes is even shorter, if the wet storage needs to be assessed. It is therefore of paramount importance to continously assess the consequences of advanced fuel assemblies on their compatibility with the different steps of the back end of the fuel cycle.

3. THE BACK END OF FUEL CYCLE STRATEGIES

The burnup increase of spent fuel is linked with an increase in dose rate and decay heat dissipation. In countries where back-end installation beyond storage in pools at reactor site (e.g. transportation system, interim storage, reprocessing) are available, these consequences have to be examined. The main steps in the back-end of the fuel cycle which are effected are

- interim storage ARS (at reactor site in the fuel pool)
- transportation
- interim or long term storage AFR (away from reactor e.g. in casks)
- reprocessing
- fuel conditioning and final disposal

In general, the effects can be reduced by extended storage on-site in the waterpools prior to transportation.

Because of the good cooling capacity of the water there are no problems from this point of view and it is advantageous to use this storage capability to the maximum extent. Also transport casks can fulfill the requirements. However, dual purpose casks do have a high pay load to make them competitive, and thus dose rate on the surface of the cask and heat dissipation do need special assessments.

Increasing decay heat dissipation leads to higher temperature of the cladding, higher EOL (end of life) internal gas pressure and increasing implication for the mechanical strain. Therefore, the storage temperatures under dry conditions are limited (to about 400 °C). Under long term storage conditions the history of the temperature gradient is also important.

3.1   Interim storage at reactor site

In most cases interim storage ARS occurs in water filled storage ponds. Figure 2 compiles all essential mechanisms affecting spent fuel cladding performance during wet storage [4, 5].

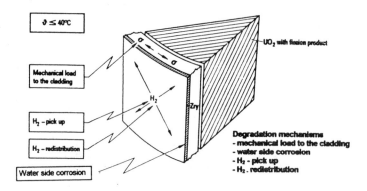

Fig. 2   Mechanisms affecting spent fuel cladding performance during wet storage

Assessment of the ability of fuel assemblies to be safely held in wet stores - in particular for long periods - has been based above all on the evaluation of corrosion mechanisms: oxidative and electrochemical cladding-tube corrosion, corrosive attack of structural components and so-called crevice corrosion. Pure oxidative corrosion is of no importance. Electrochemical attack can be suppressed through the selection of appropriate materials and by adequate control of the pool water chemistry. Furthermore, fuel assemblies are for the most part passivated after irradiation in the reactor. All defect mechanisms associated with stress and strain in the Zircaloy claddings can be neglected since these stresses are far less than the yield strength.

The major mechanisms affecting the spent FA storage performance are stress driven. Higher burnup increases the internal fission gas pressure and reduces the cladding wall thickness by oxidative corrosion. But due to the low wet storage temperature the stress still remains below the yield strength consequently wet spent fuel storage is of no concern when increasing the burnup.

3.2 Long term dry storage

Figure 3 features a compilation of all noteworthy potential degradation mechanisms [4, 5].

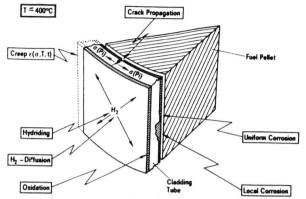

Fig. 3 Mechanisms affecting spent fuel cladding performance during dry storage

For safety reasons, spent fuel should remain intact during dry storage. Creep does not damage Zircaloy cladding if the total strain during the storage period remains below 2 to 3 %. Zircaloy oxidation is a thermally-induced process. If inert gas is employed for the storage atmosphere, no problems will arise. Oxidizing atmospheres, however, require lower storage temperatures. Iodine stress corrosion cracking needs a specific temperature range, but chemically active iodine will scarcely be available due to the chemical conditions inside the spent fuel rod. Fuel rods that have sustained damage during reactor operation will not experience defect propagation in an inert gas atmosphere. Oxidizing storage e.g. air atmospheres do, however, require further investigation.

The situation arising from higher burnup levels is much more complex for dry storage than for wet storage, for the following three reasons:

- storage temperatures are higher under dry conditions
- the fuel rod internal gas pressure is higher
- the residual wall thickness of the cladding can be smaller.

Typically the higher dry storage temperatures and the higher fission gas pressures originate from the higher end-of-life (EOL) burnup and from a higher average gas temperature in the fuel rod at the beginning of storage. This will generate higher stresses and strain in the cladding. Since a longer residence time of the fuel assemblies in the core also tend to decrease the residual metal wall thickness through corrosion of the cladding material, stress and strain in the cladding is increased. The higher storage temperatures of dry stores - which depend on the storage technology applied - serve to promote all thermally-induced degradation processes.

Therefore, the loading schemes for transport and storage casks need to be defined in more detail in the future in order to fulfill all licensing requirements based on specific EOL-conditions and all other relevant FA-criteria. One of the major topics is the computation of cask loading schemes taking into account the available FA with their specific n-and y-emissions in order to load the best possible

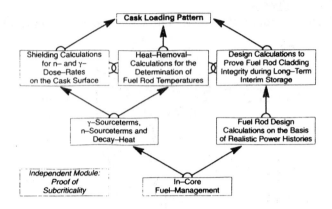

Fig. 4 Procedure to generate optimized cask loading patterns

selection of various FA into the cask (Fig. 4). The resulting time-dependent FA temperatures will cause time-dependent fuel rod cladding strain. An appropriate cask loading scheme guarantees, that these strains stay within the limits of the license.

The potential benefit of such a procedure is demonstrated by a comparison of a calculated and measured γ-dose rate at the outside of a TN 1300 dual purpose cask. The demonstration experiment was performed in 1983 using Biblis FA. The calculation of the source terms was based on EOL data reflecting the real individual burnup history from each FA. Figure 5 demonstrates the excellent coincidence of measured and calculated dose rate.

n-dose rate

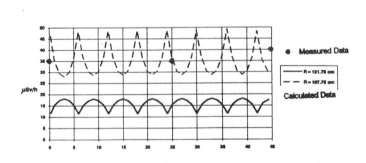

Fig. 5 Dose rate of a TN 1300 cask: measured and calculated data

t should be remembered that calculated source term data and surface γ- and n-dose rate often deviates considerably from measured data if envelope data for a pent fuel batch are used, with the consequence that largely conservative source erm data are to be used for shielding and temperature calculations.

ig. 6 shows the results of the calculation from the temperature distribution in the TN 1300 cask using the correct source terms determined as described above. The measured and the calculated temperatures deviate less than 5 %. Improved predictions of spent fuel assembly temperatures in storage allows for a better description of fuel assembly storage performance.

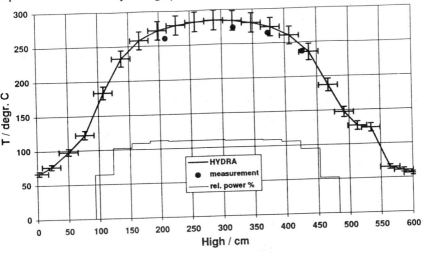

Fig. 6 Axial temperature profile of a TN 1300 cask: measured and calculated data

Replacing spent FA batch data for clad corrosion and internal fission gas pressure by individual FA-EOL data reflecting FA specific in service performance, will decrease the calculated hoop stress and related strains in dry storage considerably. Also the FA hot spot temperature will decrease, since the source terms are replaced by less conservative data as discussed above.

Fuel assembly performance prediction based on statistical methods provides for a given cask more storage capabilities in relation to fuel assemblies with a shorter pre-storage time period in water or with a higher burn-up. Fig. 7 shows a typical fuel assembly storage temperature in relation to storage time period. Fig. 8 compares the result of a deterministic and of a statistical strain calculation. Whereas the deterministic calculation results in a total strain of a little more than 2 % is the 95/99-strain - based on about 10^4 individual calculations - 1 % less.

The calculations for several variations of loading pattern based on individual FA characteristics may generate on optimized loading pattern. Taking all described benefits into account, a given cask with a given set of licensing criteria will take spent fuel with higher burnup, shorter decay time or even both without any hardware changes. Therefor it can be stated that individual FA (EOL) related design and safety criteria assessment provides increased technical capabilities for a given cask or any other confinement.

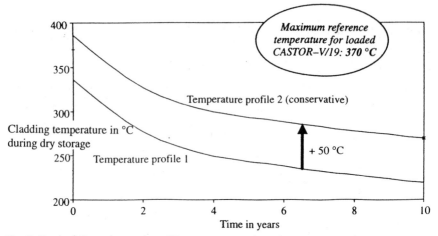

Fig. 7 Typical time dependent FA temperature in a dry storage cask

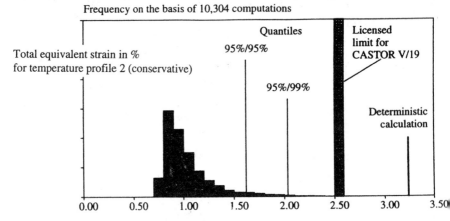

Fig. 8 Hoop strains of a FR cladding in storage: comparison of results from deterministic and statistical assessments.

3.3 Early Spent Fuel Encapsulation

A quite different approach to overcome the challenge related to increasing burnup is an early encapsulation. Spent FAs are encapsulated individually at the end of the wet storage period [6]. The encapsulation process will take place in the spent fuel pool. Preferentially it will be combined with the loading procedures of transport and/or storage casks. The FA-capsule will take over fission product retention in order to relieve the fuel rod cladding of this function. This approach will facilitate the implementation of further increasing FA burnups in future. Early encapsulation includes the possibility of avoiding any late FA-conditioning, if the spent FA-capsule is to be used also for final disposal together with an adequate overpack.Transferring the function of a fission product barrier from the cladding to the stainless steel capsule provides several advantages e.g.:

all mechanical and thermal impacts are to the adequately designed capsule thus decoupling the front end and the back requirements of the fuel cycle necessary optimization at the front end are no longer influencing the back end of the fuel cycle e.g. burnup increase.

However the fission product barrier function requires a high degree of quality for the capsule closure welding. The closure welding process is to be operated remotely within an encapsulation station placed temporarily into a fuel pond to provide the encapsulation services. An experimental study on different welding technologies resulted in the conclusion that a high energy, laser weld provides a good result for the following reasons:

- the welding process does not need any material additions
- through-going welds in several thicknesses can be realized reliably with acceptable welding speeds
- the welding process is scarcely affected by the wear of electrodes or irregularities of any feed systems

Figure 9 exhibits a typical laser weld of a capsule closure.

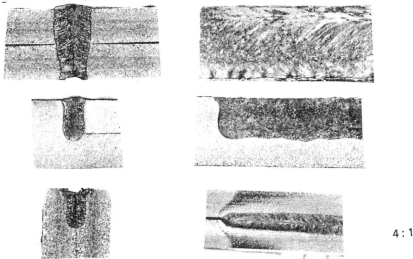

4:1

Fig. 9 Micrographic cross-section of a NdYAG laser welded capsule closure

4. CONCLUSIONS

The burnup of spent LWR fuel assemblies (FA) is growing in recent years in order to improve fuel cycle costs as a whole. Nowadays spent fuel unloaded has batch average burnups up to 50 GWd/tU. Reloads presently contracted are targetting up to 55 GWd/tU batch average. Therefore end-of life-(EOL] conditions of FA with a higher burnup need to be considered in detail in all steps of the back end of the fuel cycle.

With respect to wet fuel storage performance no major consequences are expected for the storage of spent high burnup fuel, since the stresses are still far below the yield strength in pool storage.

The situation arising from the dry storage of higher burnup fuels is much more complex. The FA-temperatures as well as the fuel rod internal gas pressure (EOL] are higher compared to wet storage. Therefore, the loading schemes of given transport and storage casks need to be defined in more detail in the future in order to fulfill all licensing requirements based on specific EOL-conditions and all other relevant FA-criteria. One of the major topics in the computation of cask loading schemes is taking into account the available FA with their specific n- and γ-emissions in order to load the best possible selection of various FA into the cask. The resulting time-dependent FA temperatures will cause time-dependent fuel rod cladding strain. An appropriate cask loading scheme guarantees that these strains stay within the limits of the license.

A quite different approach to overcome with increasing burnup is an early encapsulation of the spent FA at the end of the wet storage period. A key technology for spent fuel encapsulation is the capsule closure by remote welding. Laser welding was operated successfully.

Early encapsulation also includes the possibility of avoiding any late FA-conditioning if the spent FA-capsule is to be used for final disposal.

REFERENCES

[1]   IAEA Yearbook 1994, September 94 ISBN 92-0-102394-4

[2]   J. Banck, M. Peehs "Handling, Storage and Final Treatment of Spent FA with Increased Burnup" 5. Intern. Conf. on Rod Waste Management and Environm., Rem. Berlin, 9/95

[3]   J. Banck, M. Peehs "Long term storage in air cooled vaults in view of fuel assemblies with increased heat dissipation", Intern. Conf. of Emerging Nuclear Fuel Cycle Systems, Global 1995

[4]   M. Peehs, J. Banck, "Spent Fuel Storage: A Reliable Technology in the Back End of the Fuel Cycle", Proc. of 1993, Intern. Conf. on Nuclear Waste Management and Environmental Remediation vol. 1, p. 259, Prag, Sept. 93

[5]   M. Peehs, F. Takats, E. Vitikainen, K. Wasywich " The IAEA Coordinated Research Program on the Behavior of Spent Fuel and Storage Facility Components during Long Term Dry Storage (BEFAST), Int. Symposium on Spent Fuel Storage, Safety, Engineering and Environmental Aspects, Vienna, Okt. 1994

[6]   M. Peehs, J. Banck, V. Dannert, "Early Spent Fuel Encapsulation for Decoupling FA Operational Requirement and Long Term Dry Storage", Intern. Symp. on Spent Fuel Storage, Safety Engineering and Environmental Aspects, Vienna, October 1994

ent nuclear behaviour in long-term dry storage

WART BS, MBA
partment of Energy, Washington DC, USA
CKINNON PhD, MASME
Northwest Laboratory, Washington State, USA

SYNOPSIS

Information on spent-fuel integrity is of interest in evaluating the impact of long-term dry storage on the behaviour of spent fuel rods. Spent fuel used during cask performance tests at Idaho National Engineering Laboratory (INEL) offers significant opportunities for confirmation of the benign nature of long-term dry storage. The cask performance tests conducted at INEL between 1984 and 1991 included visual observation and ultrasonic examination of the condition of the cladding, fuel rods, and fuel assembly hardware before dry storage and consolidation of the fuel; and a qualitative determination of the effect of dry storage and fuel consolidation on fission gas release from the spent-fuel rods. A variety of cover gases and cask orientations were used during the cask performance tests. Cover gases included vacuum, nitrogen, and helium. The nitrogen and helium backfills were sampled and analysed to detect leaking spent fuel rods. At the conclusion of each performance test, periodic gas sampling was conducted on each cask as part of a cask surveillance and monitoring activity. A spent-fuel behaviour project (i.e., enhanced surveillance, monitoring, and gas-sampling activities) were initiated for intact fuel in a CASTOR V/21 cask and for consolidated fuel in a VSC-17 cask. The cask performance tests and results of the spent-fuel behaviour studies reported in this paper are based on ongoing programs by the U.S. Department of Energy (DOE).

INTRODUCTION

In response to the Nuclear Waste Policy Act of 1982, DOE issued a Solicitation for Cooperative Agreement Proposal in May 1983 to stimulate the development of dry spent nuclear fuel storage technologies. Virginia Power (VP) proposed that pressurized water reactor (PWR) spent-fuel storage cask performance testing be conducted at a Federal site in support of its at-reactor license demonstration. VP and DOE signed a Cooperative Agreement in March 1984, and VP signed a separate agreement with the Electric Power Research Institute (EPRI), essentially establishing a three-party cooperative agreement.

The scope of the Cooperative Agreement included performance testing of three different metal storage casks loaded with unconsolidated spent nuclear fuel. The tests were conducted at INEL with the GNS CASTOR V/21, Transnuclear TN-24P, and Westinghouse MC-10 casks. Prior to testing, the VP Surry reactors PWR spent-fuel assemblies used in the cask performance tests were characterized using in spent-fuel pool ultrasonic examinations and video scans. Cask internal cover gas samples were taken during testing. After testing, selected fuel assemblies were videotaped and photographed.

Upon completion of cask performance testing with unconsolidated fuel under the VP/DOE cooperative program, a decision was made by DOE and EPRI to extend the performance testing to include consolidated fuel in the Transnuclear TN-24P cask. As a result, a dry-rod consolidation project was also conducted at INEL. The fuel assemblies used in the TN-24P and MC-10 cask performance tests and spent-fuel assemblies from the Florida Power and Light Turkey Point reactor were consolidated and loaded into the TN-24P cask for a performance test. Later, a cooperative agreement was established with Sierra Nuclear Corporation, and 17 of the consolidated fuel canisters from the TN-24P cask were used in a performance test of the Sierra ventilated concrete cask, VSC-17. Performance test runs involved a combination of cover gases and cask orientations. The backfill environments used were vacuum, nitrogen, and helium; nitrogen and helium were sampled and analysed to detect leaking spent-fuel rods. The integrity of the fuel assemblies was determined from cover gas sampling and reported in technical conference papers starting in 1986 up to the most recent in 1993 (1). At the conclusion of each performance test, periodic gas sampling was conducted on each cask as part of a cask surveillance and monitoring activity.

This report combines the gas-sampling information from the cask performance test and cask monitoring activities. It documents the condition of the fuel from the Surry reactor prior to testing and the effect of testing on fuel integrity ascertained through gas sampling during cask performance tests at INEL using both intact and consolidated PWR spent fuel. It also includes results of a prior testing using boiling water reactor (BWR) spent fuel and the REA-2023 cask at Morris, Illinois. The pretest condition of the fuel is described plus the significant results obtained from gas sampling during and after performance testing. Recent gas sampling data associated with cask surveillance and monitoring at INEL is also included.

Three types of spent fuel and five casks have been used during the cask performance testing and demonstration program. General Electric 7x7 BWR spent-fuel assemblies from the Nebraska Power Cooper reactor were used for the performance test of the REA-2023 cask. Westinghouse 15x15 PWR was used in the CASTOR V/21, TN-24P, MC-10, and NUHOMS performance tests. A portion of this fuel was consolidated at INEL and used in performance tests of the TN-24P and VSC-17 casks, also at INEL. Prior to cask performance testing at GE-Morris, calorimetry testing was performed on the BWR spent-fuel assemblies. At the conclusion of testing activities, the BWR assemblies were returned to the spent-fuel pool at Morris. The PWR spent-fuel assemblies remain in dry storage at INEL. Table 1 presents a summary of the fuel used in each of the performance tests. A description of the fuel is contained in the performance reports.

Table 1. Spent-Fuel Assembly Characteristics

Cask	REA-2023	CASTOR V/21	TN-24P	TN-24P(a)	MC-10	VSC-17(a)
Fuel Type	BWR	PWR	PWR	PWR	PWR	PWR
Assembly Type	7x7	15x15	15x15	Consolidated 15x15	15x15	Consolidated 15x15
Burnup, Gwd/MTU	24-28	24-35	29-32	24-35	24-35	26-35
Cooling times, years	2.3-3.4	2.2-3.8	4.2	6.2-12.2	4.6-10.1	8.1-14
Discharged in	1981-1982	1981-1983	1981	1975-1981	1975-1981	1976-1981
Enrichment, wt%	2.5	2.9-3.1	2.9-3.2	1.9-3.2	1.9-3.2	2.56-3.2
Assy. Decay Heat, Watts	235-370	1000-1800	832-919	701-1185	400-700	700-1050
Average, Watts	290	1350	860	970	530	877
Cask, kWatts	15.2	28.4	20.6	23.3	12.6	14.9

(a) Performance test utilizing consolidated fuel in the cask.

SPENT FUEL INTEGRITY

Pretest Fuel Inspection. Four examination methods were used to assess the integrity of the spent fuel used in the cask performance tests. Methods common to the PWR and BWR fuel included visual observations, including full-length black and white videos and colour photographs; and analyses of the cover gas in the cask. In addition to these methods, the BWR spent fuel was examined by in-basin sipping and the PWR spent fuel from the VP's Surry Reactor was examined using an in-pool ultrasonic examination.

In Fuel Pool. In-pool sipping consisted of placing a hood over the selected BWR assembly and analysing the water that was drawn off the top of the assembly. All the sipping data was compared to background readings to assess fuel integrity. Although there is some variation in the differences between the pretest and post-test radionuclide concentrations, the values were lower than would exist if the assembly contained leaking fuel rods. The sipping results did not indicate any leaking fuel rods in any of the fuel assemblies used in the cask either before or after cask testing.

In Fuel Pool Ultrasonic Inspections. Ultrasonic inspections were performed on the PWR fuel at VP's Surry Reactor using the Babcock & Wilcox Failed Fuel Rod Detection System (FRDS). The FRDS system uses ultrasonic techniques to differentiate between non-leaking and leaking rods by detecting the presence of moisture in the latter. Only Surry Reactor fuel assemblies with non-leaking fuel rods were used in the performance tests.

Visual, Video and Photographic Examinations. The PWR fuel assemblies were examined visually to establish their general condition after shipment from VP, after handling at the INEL Hot Shop, after cask performance testing, and during consolidation. Similar exams were made of the Cooper BWR fuel during the REA-2023 performance tests at GE-Morris. Two kinds of visual examinations were used: black-and-white videos and colour photography of selected fuel assemblies.

The black-and-white videos taken at GE-Morris, VP, and INEL did not provide sufficient detail to characterize the crud or very small features on the fuel rods. They did not reveal any indication of significant variations in the fuel rods after shipment, handling, and performance testing. The resolution of the videotapes did not provide enough information to adequately determine the integrity and condition of the fuel and fuel cladding. Examination of the video scans showed that all the fuel assemblies and fuel rods look basically the same when viewed from outside the assemblies. There was some discoloration of the fuel rod cladding in the area of the grid spacers, which was expected.

Colour photographs showed that a typical orange/reddish crud (probably Fe_2O_3) was evenly deposited on all of the Zircaloy 2 cladding and fuel assembly hardware. There were no noticeable changes in the characteristics or adherence of the crud during handling operations involving the spent-fuel assemblies at GE-Morris or INEL. Some scratches and worn spots were apparent on the spacer grids and some fuel rods, but these features did not change as a result of examination or handling operations. In general, the fuel rods were in excellent condition with a very adherent crud layer.

Additional visual examinations of the fuel were conducted during the dry-rod consolidation program. According to Vinjamuri (2 and 3):

"No noticeable cladding defects in the rod surfaces were observed for any of the fuel processed. The oxide layer on the surface of the fuel rods appears to be intact and firmly attached to the cladding. The oxide layer does not appear to be loose, thick, soft, or powdery. However, the oxide layer and some of the zirconium cladding was scraped from the rod surface by the spacer grids as the rod was pulled during fuel consolidation. Very little crud buildup on the surfaces of the rods was observed. The surfaces of the rods displayed only a thin oxide layer, which had the appearance of surface discoloration rather than any rough or loose material. The rod surfaces are discoloured near the spacer grids. The discoloration has an appearance of a dark mottling of the surface and is progressively more predominant from the middle of the rod length toward the rod bottom. The rods are generally clean, with limited amounts of clad discoloration and oxidation... The evidence of fuel rod growth since fabrication was visually obvious during the consolidation process... Length variation between rods appears to be as much as 2 cm (0.8 inch). The rods that grew longer than others appeared to be randomly located within the fuel assembly."

Cask-Cover Gas Sampling. The cask-cover gas was sampled several times during each cask performance test to evaluate the integrity of the spent-fuel rods. Each sample was collected in a separate 500-cc stainless steel cylinder. The cylinders were checked for leaks before sampling. Initially, during the CASTOR-V/21 cask performance test, the cylinders were equipped only with quick disconnect-fittings and no bellows-sealed valves as part of the closure. During the early sampling efforts with the CASTOR-V/21 cask, the cover gas samples in the cylinders were diluted with ambient air from the vicinity of the sampling apparatus, air that leaked into the cylinder during shipment, and argon introduced at Lawrence Livermore National Laboratory where some of the samples were analysed. In many cases, this dilution was made more severe by the collection of small amounts of cask cover gas, presumably due to short equilibration times between the cask and the sample bottle during the actual cask cover gas collection procedure. The end effect of small, diluted samples on the cask cover gas analyses was to increase detection limits, increase measurement uncertainties, and introduce questions of sample validity. Once bellows-sealed valves were added to the sampling cylinders, the problem of air leakage into the sampling cylinders was eliminated.

Gas sample analysis included mass spectroscopy and radiochemical gamma analysis. Mass spectra were analysed for all common fixed gases with masses less than 100 to verify the purity of backfill gas composition. Only N_2, O_2, He, Ar, and CO_2 concentrations above 0.01 percent are detected in any of the samples. The integrity of the fuel rods was assessed from radionuclide concentration based on gamma spectroscopy.

Radiochemical gamma analysis was used to detect Krypton-85. The relatively low amounts of Krypton-85 detected indicate that no leaking fuel rods were present in the GNS CASTOR-V/21 and MC-10 casks during performance testing with unconsolidated fuel and up to about a year after testing. At this time gas sampling in these casks was discontinued. The final gas sample taken from the CASTOR-V/21 cask during this period of time was taken in December 1986. In September 1994, the CASTOR cask was opened and backfilled with a fresh charge of helium gas. The pre and post test backfill was checked for purity. Gas samples taken in March and July 1995 (after six and nine months of gas residence in the cask) did not contain detectable amounts of Krypton-85 that would indicate leaks from fuel rods during that storage period. This is particularly significant because the first few assemblies loaded in the CASTOR-V/21 cask were exposed to air for approximately 200 hours during incremental loading of the cask and fuel assembly/basket inspections at a reduced temperature. In addition, after testing was completed and long-term surveillance started, all the fuel assemblies were in a 70-percent He and 30-percent air environment for approximately 4 months because a quick-disconnect fitting on the CASTOR-V/21 cask lid had not sealed properly.

Two casks loaded with intact fuel have shown Krypton gas concentrations indicative of a leaking fuel rod. During the performance test of the REA-2023 cask, Krypton gas was detected in the cask after being rotated from a vertical to horizontal orientation. The accumulated amount of Krypton gas released to the cask was consistent with the release from a single fuel rod (4 and 5). The cladding defect was assumed to be very small since the release rate was essentially linear during 2.5 months of testing. There was no confirmation of a leaking fuel rod either by visual inspection or sipping of the fuel assemblies after the cask test. The gas analyses provided the only indication of a leaking fuel rod. The leaking fuel rod had no impact on the basin operation or handling of the fuel assemblies subsequent to the cask test.

The other cask, which was loaded with intact fuel and showed Krypton levels indicative of a leaking fuel rod, was the TN-24P cask. In this performance test, the cumulative amount of Krypton-85 detected just after the cask was rotated from a vertical to a horizontal orientation indicated a fuel rod leaked during this portion of the test. The decay in the leak rate, as indicated by subsequent gas samples, indicates that the leak was small. It took several days to vent the gas from the fuel rod.

In May 1987, 36 of the 48 intact fuel assemblies in the TN-24P and MC-10 casks, plus 12 intact assemblies that had been in the Turkey Point Reactor, were consolidated into 24 consolidated fuel canisters as part of INEL's Dry Rod Consolidation Technology Project. The consolidated fuel canisters were then used in performance tests of the TN-24P and VSC-17 casks. During the fuel-rod consolidation process, the exhaust gases from the consolidation area were monitored to detect the release of radioactive gases from the fuel that would indicate a cladding failure. In the consolidation reports (2 and 3), one of the conclusions reached was that all fuel rods from the 48 assemblies were pulled and canisterized without rod failures.

Later, during the performance test of the TN-24P cask using consolidated fuel, Krypton-85 was released to the cask. Based on a combination of ORIGEN2 predictions and experimental measurements (4 and 5), it was estimated that four or more fuel rods may have developed leaks between the end of cask loading and the beginning of cask performance testing, three or more fuel rods during cask performance testing, and another five fuel rods in the six-month period following testing. The rate of Krypton-85 release was observed to decrease with time from cask loading. Shortly after the last gas sample was taken from the fully loaded TN-24P cask, 17 canisters of consolidated fuel were removed from the TN-24P cask and loaded into the VSC-17 cask. The performance tests for the VSC-17 cask showed a nominal amount of Krypton-85 but not enough to indicate a new leaking fuel rod. Since the end of the VSC-17 performance testing in early 1991 until September 1994, the VSC-17 has been undisturbed. Recent gas samples, taken since September 1994, indicate that the atmosphere in the VSC-17 has not changed significantly. There has been a small amount of Krypton-85 release, below

© IMechE 1996 C512

the quantity expected for a single rod release, and there has been buildup of hydrogen in the cask. The amount of hydrogen is consistent with off-gassing of the RX277 neutron shield material in the lid. Similar amounts of hydrogen were observed during cask performance testing.

The amount of Krypton-85 released during and after the TN-24P cask performance test with consolidated fuel is significantly higher than that released in previous cask testing with unconsolidated fuel. Before this test, four cask performance tests of similar duration and scope had been performed; only two indications of Krypton-85 release were observed. The magnitude of the releases in the previous tests and surveillance periods indicated that each was limited to a single rod-cladding breach. The previous tests involved about 16,700 spent fuel rods, whereas this test involved about 9,800 rods. It is hypothesized that the greater magnitude of Krypton-85 released in this test and post-test surveillance is because of additional cladding leaks caused by enlargement of incipient cladding flaws during pulling and flexing of the fuel rods during the consolidation process. The enlarged cladding flaws combined with cladding creep during cask testing and surveillance periods allowed leak paths to develop. The leakage has not affected operations.

SPENT-FUEL BEHAVIOUR PROJECT RESULTS

Currently, the CASTOR V/21 cask is fully loaded with 21 fuel assemblies and has been undisturbed since July 1985 when temperatures in the cask were about 350°C; the MC-10 cask has capacity for 24 assemblies and contains 18; the TN-24P cask has capacity for 24 assemblies and contains 7 consolidated fuel canisters; and the VSC-7 cask has been fully loaded with 17 consolidated canisters since September 1990 when the temperature in the cask was about 320°C. Management of the storage casks deployed at INEL includes routine monitoring, cask and equipment maintenance, and spent-fuel behaviour studies. Routine monitoring of the casks includes visual surveillance, monitoring gas pressure between the seals or inside the casks cavity, monitoring radiation fields around the casks, and periodic sampling of the soil around the cask storage pad.

Two casks, the CASTOR V-2 1 and VSC-17, are used for an enhanced effort known as the Spent-Fuel Behaviour (SFB) Project. The selection of casks for this enhanced monitoring work was based on the initial storage temperatures and duration of storage. As previously mentioned, the spent fuel in the CASTOR V/21 cask has been undisturbed since July 1985. The VSC-17 cask contains consolidated fuel rods from 2 spent-fuel assemblies that were placed in 17 separate canisters for a total inventory of 34 assemblies. These casks represent the hottest commercial spent-fuel dry storage inventories at the INEL facility with initial storage temperatures of 320°C–350°C. Spent-fuel load patterns for the GNS CASTOR V/21 and VSC-17 casks are shown in Figures 1 and 2 (6).

The SFB Project provides information on the performance of the spent nuclear fuel and casks through increased monitoring, inspection, and sampling. The objective of the SFB Project is to confirm the predicted behaviour of spent nuclear fuel in long-term interim dry storage at Federal and commercial facilities. Specific activities include monitoring, gas sampling, potential fuel inspection, and records maintenance. These activities are listed below:

- Monitoring. Routine monitoring to comply with ongoing environmental, safety, and health considerations.

- Gas sampling. Gas samples (500 cc) are taken from each of the two casks and analysed for sample pressure, mass spectra, and radionuclide concentration to ensure backfill purity and fuel-cladding integrity.

- Fuel inspection. Additional fuel inspection will be initiated when any of the following conditions are encountered:

  - Gas-sampling analysis results show sufficient concentrations of Krypton-85 and/or Carbon-14 to conclude that there has been a fuel-cladding breach.

  - A major transfer for the cask and/or spent fuel is planned.

  - A leak is detected in the cask seal system requiring removal of the cask closure lid for repair.

- Fuel inspection records maintenance. Specific fuel inspection records and documentation will include:

  - Video coverage of spent-fuel loading/unloading.

  - Photographing of all four sides of the spent-fuel assembly using periscope camera to show crud conditions.

  - Fuel assembly smears taken remotely while fuel is out of the basket to measure how friable the crud is. Smears taken of the basket slot to measure the cask internal contamination.

  - Cask fuel radiation-level profiles.

  - Cask internal condition inspection for basket cracks, warping, discolorations, etc.

  - Prior to lid closure, seal conditions and surfaces checked and replaced as needed.

Results of the initial gas-sample analysis of the VSC-17 and CASTOR V/21 casks for the first, second, and third quarters of Fiscal Year 1995 are shown in Tables 2, 3, and 4 (7). Spectroscopy and Carbon-14 analyses of cask-cover gas samples over the past 12 months found no indication or evidence of fuel-rod cladding failure. Based on the analyses, it was concluded that no indication of fuel-rod failure was present in either cask, and inspection of the spent fuel and cask internals is not justified at this time.

SUMMARY AND CONCLUSIONS

Radiochemical gamma analysis of gas samples from cask performance tests and subsequent cask surveillance and monitoring activities provide an indication for spent-fuel integrity during dry storage. The gas-sampling analysis shown indicates that dry storage of spent fuel in an inert atmosphere is benign. In general, fuel-handling activities have a more significant impact on fuel rods than does extended dry storage in an inert atmosphere.

As a result of the Department's cooperative programs with electric utilities and industry, seven dry-storage spent nuclear fuel casks have been approved by the Nuclear Regulator Commission with 20-year certificates of compliance as follows:

- GNS Castor V/21, Certificate No. 1000
- Westinghouse MC-10, Certificate No. 1001
- Nuclear Assurance Corporation, NAC-S/T, Certificate No. 1002
- Nuclear Assurance Corporation, NAC-C.28-S/T, Certificate No. 1003
- VECTRA/NUHOMS - 24P and 52B, Certificate No. 1004
- TN-24, Certificate No. 1005
- Pacific Sierra Nuclear - VSC-24, Certificate No. 1007

In addition, the Castor X-33 was loaded at the Surry facility of Virginia Power in February 1995, and the VECTRA MP-187 is undergoing licensing activities for use as a transportable storage system. It is expected that data from the SFB Project will be essential for Nuclear Regulatory Commission recertification of dry-storage casks beyond their initial 20-year term.

REFERENCES

1. McKinnon, M.A., and V.A. Deloach, 1993. Spent Nuclear Fuel Storage - Performance Tests and Demonstrations, PNL-8451, Pacific Northwest Laboratory, Richland, Washington.

2. Vinjamuri, K., E.M. Feldman, C.K. Mullen, A.E. Arave, B.L. Griebenow, J.H. Browder, P.E. Randolph, and W.J. Mings, 1988a. Dry Rod Consolidation Technology Project Quick-Look Report, EGG-WM-7852, EG&G Idaho, Idaho Falls, Idaho.

3. Vinjamuri, K., E.M. Feldman, C.K. Mullen, B.L. Griebenow, A.E. Arave, and R.C. Hill, 1988b. Dry Rod Consolidation Technology Project at the Idaho National Engineering Laboratory, EGG-WM-8059, EG&G Idaho, Idaho Falls, Idaho.

4. Barrier, JO., 1985. Characterization of LWR Spent Fuel MCC - Approved Testing Material--ATM-10l, PNL-5109, Pacific Northwest Laboratory, Richland, Washington.

5. Guenther, R.J., D.E. Blahnik, T.K. Campbell, U.P. Jenquin, J.E. Mendel, L.E. Thomas, and C.K. Thornhill, I 988. Characterization of Spent Fuel Approved Testing Material--ATM-103, PNL-5 109-103, Pacific Northwest Laboratory, Richland, Washington.

6. EG&G, June 1994. Monitoring Plan - Spent Fuel Behaviour in Long-Term Dry Storage.

7. Lockheed Idaho Technologies Company. Spent Fuel Behaviour in Long-Term Storage Project Fiscal Year 1995 Quarterly Reports.

Figure 1. GNS Castor V/21 Cask Spent Fuel Load Pattern

The Gesellschaft fur Nuklear Service (GNS) Castor V/21 cask contains 21 PWR Westinghouse 15x15 spent fuel assemblies.

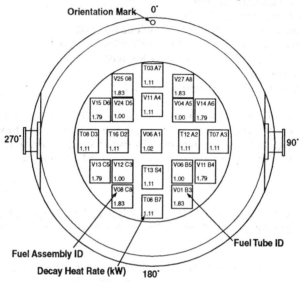

Figure 2. Sierra Nuclear VSC-17 Spent Fuel Load Pattern

The VSC-17 cask contains 17 canisters of consolidated fuel rod taken from 34 PWR Westinghouse 15x15 spent fuel assemblies

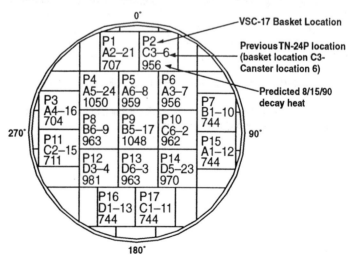

Table 2. 1st Quarter FY95 Gas Sample Data Summary

(mole percentages except where stated)

GAS	INEL CPP LAB		ARGONNE-WEST LAB		INEL CPP LAB		ARGONNE-WEST LAB	
	GNS - 1	GNS - 2	GNS - 3	GNS - 4	VSC - 1	VSC - 2	VSC - 3	VSC - 4
H_2	0.04	0.04	0.04	0.04	2.0	2.3	2.3	2.5
H_e	97.75	97.9	97.8	97.9	85.2	97.0	91.0	97.0
N_2	1.77	1.83	1.72	1.69	10.0	0.55	5.3	0.48
O_2	0.41	0.43	0.38	0.36	2.62	0.10	1.3	0.07
A_r	0.02	0.02	0.02	0.02	0.12	<0.01	0.07	D-N/Q
CO_2	<0.01	<0.01	<100ppm	<100ppm	0.01	<0.01	D-N/Q	<100ppm
K_r	ND	ND	<100ppm	<100ppm	ND	ND	<100ppm	<100ppm
X_e	ND	ND	<100ppm	<100ppm	ND	ND	<100ppm	<100ppm

(micro-Curies per cubic centimeter)

GAS	INEL LAB		ARGONNE-WEST LAB		INEL LAB		ARGONNE-WEST LAB	
	GNS - 1	GNS - 2	GNS - 3	GNS - 4	VSC - 1	VSC - 2	VSC - 3	VSC - 4
C-14	—	—	—	—	—	—	—	—
Co 60	—	—	—	Detected[1]	—	—	—	Detected[1]
Kr 85	9.575.E-03	4.727.E-03	ND	ND	—	—	1.70E-02	1.26E-02
Cs 137	—	—	Detected[2]	Detected[2]	—	—	Detected	Detected

1 - Activity not measured from calibrated samples

Table 3. 2nd Quarter FY95 Gas Sample Data Summary

(mole percentages except where stated)

GAS	INEL CPP LAB		ARGONNE-WEST LAB		INEL CPP LAB		ARGONNE-WEST LAB	
	GNS - 1	GNS - 2	GNS - 3	GNS - 4	VSC - 1	VSC - 2	VSC - 3	VSC - 4
H_2	< 0.01	< 0.01	< 0.01	< 0.01	2.09	2.09	2.11	2.12
H_e	99.43	99.28	99.37	99.42	96.96	97.75	95.51	96.12
N_2	0.56	0.67	0.60	0.56	0.76	0.14	1.91	1.41
O_2	<0.01	0.03	0.02	0.01	0.17	<0.01	0.44	0.32
A_r	<0.01	<0.01	<0.01	<0.01	0.01	<0.01	0.03	0.02
CO_2	<0.01	0.01	<0.01	<0.01	<0.01	<0.01	<0.01	<0.01
K_r	ND	ND	<0.01	<0.01	ND	ND	<0.01	<0.01
X_e	ND	ND	<0.01	<0.01	ND	ND	<0.01	<0.01

(micro-Curies per cubic centimeter)

GAS	INEL LAB		ARGONNE-WEST LAB		INEL LAB		ARGONNE-WEST LAB	
	GNS - 1	GNS - 2	GNS - 3	GNS - 4	VSC - 1	VSC - 2	VSC - 3	VSC - 4
C-14		5.4E-06	—	—	3.6E-06	4.1E-06	—	—
Co 60	—	—	—	—	—	—	—	—
Kr 85	6.6E-06	6.8E-06	ND	ND	6.6E-03	6.5E-03	1.03E-02	1.19E-02
Cs 137	—	—	—	—	—	—	—	—

Table 4.   3rd Quarter FY95 Gas Sample Data Summary

(mole percentages except where stated)

GAS	INEL CPP LAB		ARGONNE-WEST LAB		INEL CPP LAB		ARGONNE-WEST LAB	
	GNS - 1	GNS - 2	GNS - 3	GNS - 4	VSC - 1	VSC - 2	VSC - 3	VSC - 4
H_2	<0.01	<0.01	0.02	0.02	2.15	2.16	2.25	2.22
H_e	98.52	99.40	99.40	99.41	97.66	97.70	97.62	97.65
N_2	1.28	0.59	0.55	0.55	0.17	0.13	0.12	0.12
O_2	0.2	<0.01	0.01	0.01	0.01	<0.01	<0.01	<0.01
A_r	0.01	<0.01	<0.01	<0.01	<0.01	<0.01	<0.01	<0.01
CO_2	0.01	<0.01	0.01	0.01	<0.01	<0.01	<0.01	<0.01
K_r	ND	ND	<0.01	<0.01	ND	ND	<0.01	<0.01
X_e	ND	ND	<0.01	<0.01	ND	ND	<0.01	<0.01

(micro-Curies per cubic centimeter)

GAS	INEL LAB		ARGONNE-WEST LAB		INEL LAB		ARGONNE-WEST LAB	
	GNS - 1	GNS - 2	GNS - 3	GNS - 4	VSC - 1	VSC - 2	VSC - 3	VSC - 4
C-14	1.08E-05	1.08E-05	—	—	8.0E-07	1.1E-06	—	—
Co 60	—	—	—	—	—	—	—	—
Kr 85	4.0E-08	4.0E-08	ND	ND	6.80E-03	6.80E-03	8.28E-03	8.01E-03
Cs 137	—	—	—	—	—	—	—	—

e Swedish spent fuel encapsulation facility – design
atures

:DMAN
dish Fuel and Waste management Company, Stockholm, Sweden
CTON MA, MEng, CEng, IMechE, IMechC
h Nuclear Fuel Engineering Limited, Manchester, UK

SYNOPSIS

The Swedish Nuclear Fuel and Waste Management Company, SKB, propose to construct a facility
to encapsulate spent nuclear fuel in disposal canisters for emplacement in a deep repository. This
plant will be the subject of an application for construction to the Government in a few years. The
current programme of work includes planning and testing of encapsulation methods.

The disposal canister must be designed to remain intact for a very long time and the proposed
construction is for an insert to provide mechanical strength (e.g. Steel) and a copper overpack to
provide corrosion protection. The encapsulation process includes equipment for fitting the lid of
the copper overpack using electron beam welding and inspection equipment to verify the weld
quality.

This work is in two parts; development and fabrication of canisters and the design of an
encapsulation plant. This paper reviews the current status of the plant design and describes the
main features of the process. It also highlights those areas of the process where development is
continuing and latest plans for a Laboratory for Investigation of Encapsulation techniques.

The encapsulation plant is planned as an extension to the existing central storage facility
(CLAB). The encapsulation process is in four distinct stages; retrieval of fuel from the CLAB
storage pools, checking and drying of fuel assemblies, encapsulation in the disposal canister and
finally inspection of the completed canister. The handling processes therefore include pool

equipment similar to the CLAB equipment and remote handling techniques within hot cells from the drying stage onwards.

INTRODUCTION

The Swedish nuclear power utilities are responsible for the safe management and disposal of spent nuclear fuel and other radioactive waste from the 12 Swedish nuclear power stations. In order to fulfil this responsibility, the utilities have jointly created SKB, the Swedish Nuclear Fuel and Waste Management Co, with the task to plan, build and operate the necessary waste management facilities and systems.

BNFL Engineering Ltd have been working with SKB for the last 3 years to develop the design of the Encapsulation Plant with particular responsibility for the remote handling processes of the system.

THE FUEL STORAGE PROCESS

SKB has developed a system that ensures the safe handling of spent fuel from the Swedish nuclear power plants for the foreseeable future. The keystones of this system are:

1. A central interim storage for spent nuclear fuel and core components, CLAB, in operation since 1985.
2. An encapsulation plant for spent nuclear fuel.
3. A deep repository for encapsulated spent fuel and other long-lived radioactive wastes.

CENTRAL INTERIM STORAGE FOR SPENT FUEL (CLAB)

CLAB is located at the Oskarshamn nuclear power plant on the east coast of Sweden. Operation started in 1985, and at the end of 1995 more than 700 casks containing some 2300 tonnes of fuel had been received. Also, approximately 70 casks with activated core components, e.g. control rods, had also been received.

The main complex above ground is the receiving building, where the fuel transport casks are unloaded. The unloading is performed under water. The CLAB storage section is located in a rock cavern, the roof of which is 25-30 metres below ground level.

The present capacity of CLAB, approximately 5000 tonnes of uranium, covers the requirements until 2004. The Swedish nuclear program is expected to generate approximately 8000 tonnes of which approximately 75 percent is BWR fuel and 25 percent PWR fuel. CLAB must therefore be expanded by adding storage pools in a new rock cavern close to the existing one. According to he current plan, the construction of the second cavern will start in 1999. The spent fuel will emain in CLAB for approximately 30 years to decay before encapsulation.

THE FUEL ENCAPSULATION PLANT (FEP)

The Fuel encapsulation plant is planned to be built adjacent to the CLAB building. The plant will encapsulate. Fuel received from CLAB in copper disposal canisters after drying and conditioning of the Fuel. The disposal canisters will be sealed by fusion welding. After monitoring and decontamination they will be held in a buffer store pending export to the repository

The Basic Design of the encapsulation plant was completed in early 1996 and is the base for the Preliminary Safety Report. The licensing application is programmed for submission to the Swedish authorities in early 1998. The construction work for the encapsulation plant is expected to start at the end of the century. The commissioning operations of the facility would then start in 2005. Delivery of disposal canisters to the final deep repository is planned for 2008. The design capacity of the plant is one canister per working day, corresponding to an annual output of approximately 210 canisters. Core components and reactor internals will also be conditioned in the encapsulation plant. These activities are planned to start approximately year 2020, after evaluation of the first operation stage of the spent fuel encapsulation and disposal operation.

REPOSITORY

The completed disposal canisters will be transferred in a transport cask to a deep hard rock repository. Within the repository the canisters will be lowered into a deposition hole and packed with bentonite.

The site for the repository is currently under investigation. SKB are currently operating a deep rock laboratory at ÄSPÖ. The laboratory is used for the investigation of repository construction methods, geological investigation and development of packing materials.

Future work at the laboratory will include the construction of a prototype repository to develop and prove the canister handling and deposition technology and trials on the deposition and recovery of dummy canisters. It is planned to operate the laboratory well into the next century to support the construction and operation of the repository.

NOVEL-DESIGN FEATURES OF THE PROCESS

The concept of encapsulating fuel for disposal in a deep repository presents a number of novel design features. These are:

a) The design of a disposal canister with the required life.

b) The development of a reliable canister welding technique for the sealing of the canister.

c) The development of an efficient reliable mechanical handling process for installing the fuel within the canister.

The following sections of this paper describe the design development work completed to date on the above and describes the planned future work to complete the development of the encapsulation process.

GENERAL DESIGN REQUIREMENTS FOR THE CANISTER

The canister must be designed and fabricated in such a manner that they remain intact for a very long time in the conditions that will prevail in the deep repository. This means that they must not be penetrated by corrosion in the groundwater present in the rock, or be broken apart by the mechanical stresses to which they are subjected in the deep repository. To achieve this, the canister is planned to consist of an insert e.g. steel, which provides mechanical strength, and an overpack of copper, which provides corrosion protection. Copper corrodes very slowly in the oxygen-free water present at depth in Swedish bedrock. Studies have shown that the canister will probably remain intact for millions of years, which provides a considerable margin of safety (which is needed due to uncertainties and variations in the premises).

The requirement for the canister to be impervious can be subdivided into initial integrity, corrosion resistance and strength. The initial integrity means that the canister must be fabricated, sealed and inspected with methods that guarantee that very few (0.1%) will contain undetected defects that could entail initial leakage or that could lead to early canister failure. The canister must not be penetrated by corrosion so that water can enter the canister during the first 100,000 years after deposition. The maximum corrosion depth is estimated to be about 5mm in typical Swedish ground water. In recognition of uncertainties in the data and other factors, a suitable safety factor should be included when determining wall thickness. The canisters must be designed to withstand deposition at a depth of 400-700 m.

No temperature limit has been established for the spent fuel in storage, due to the fact that the temperature has been considerably higher during the fuels transport to the CLAB facility. However, the maximum surface temperature of the canister must be limited to 90 °C to avoid excessive heat transfer to surrounding bentonite and evaporation of groundwater outside the canister. This could lead to enrichment of for example chlorides on the canister surface.

Safety during operation and maintenance work at the encapsulation plant must be high. The canister's design must meet the requirements made by both normal and abnormal operating cases in the plant. It must also be able to withstand handling accidents that could give the personnel and the plant to unacceptable exposures or lead to unacceptable releases of radioactivity. During ransport to the deep repository, the canister is placed in a transport cask which provides rotection against external damage. For handling and disposal in the deep repository, the canister nust be designed so that it can be transferred from the transport cask to the deposition equipment. Emplacement in the deposition holes must be able to be performed with the necessary precision nd safety. In the event of retrieval of canisters after the initial deposition stage, it must be possible to lift the canisters and place them in a transport cask.

The safety goals should be fulfilled with observance of good resource management and in consideration of the environmental consequences of canister fabrication and the encapsulation procedure. The selected canister material must not have any harmful effect on the near-field nvironment.

REFERENCE CANISTER

The work with the canister and the encapsulation process has focused on studies and development of technology that does not require heating of the fuel during encapsulation. Such technology facilitates the encapsulation process and reduces the radiological risks for the operating personnel. The canister is planned to be composed of two components: an outer corrosion protection of copper and an inner pressure-bearing container of steel so that it will fill its function in the deep epository (Figure 1).

A copper canister with a 50 mm wall thickness made of oxygen-free copper with a low phosphorus content is being used as reference for the continued work. The bottom and lid are oined to the shell by electron beam welding. The canister insert is of cast steel and has a minimum thickness of 50 mm. Several alternative designs of the copper canister have been studied and will be further studied. The inner container is cast in one piece with holes for the different ypes of fuel assemblies. It is assumed that it will be cast in steel, iron or perhaps some other material. This design comprises the reference for the continued work. The exact size of the canister and choice of material grades will be studied in the continued work and be chosen with a view towards the criteria described above.

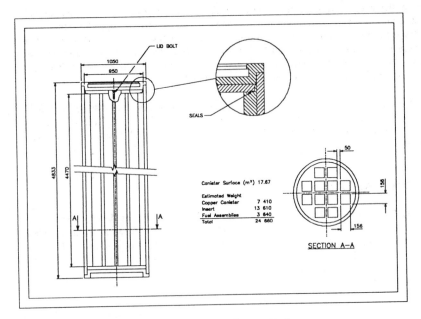

Figure 1 - Reference Canister

The choice of material in the copper canister is determined by requirements on corrosion resistance, ductility, weldability and the ability to fabricate with a suitable grain size. An oxygen-free copper with approximately 50 ppm phosphorus is being used as a reference material for the time being. The corrosion properties of this material are well known. Its susceptibility to stress corrosion cracking is low. Several possible materials are available for the inner container. The choice will be determined in part by the fabrication method.

As an alternative for the insert, a cast inner component of spheroidal graphite iron is considered. It will be identical in all essential respects to the reference alternative. Spheroidal graphite iron has better castability, but its weldability is lower. The latter does not have to be a disadvantage, since there are good prospects for casting the inner component in one piece, which would greatly reduce the requirements on weldability. A third alternative, a bronze inner component would eliminate all possibilities of hydrogen gas formation in conjunction with corrosion, if the copper shell should be penetrated. This is judged to facilitate performance assessment for a leaking canister, but the canister cost is higher for this alternative.

The size of the canister will be determined with a view to the limitations of the handling, transportation and deposition systems, and well as to the fact that the temperature on the canister surface in the deep repository must not exceed 90°C. Based on this a reference canister for up to

BWR assemblies or 4 PWR assemblies was chosen. Further studies are under way to determine e suitable canister size. They include, among other things, analyses of how sensitive the mperature is to variations in the thermal conductivity of the bentonite and the rock, rock mperature, repository configuration and deposition schedule. The preliminary data for the ference canister is 1050 mm in diameter, length 4850 mm and weight 25000 kg.

The canister must be designed so that there is no risk of criticality in connection with handling the canister with fuel in the encapsulation plant or in the long run in the final repository if water ould enter the canister. With the enrichments that are used today, a critical configuration can be hieved if the fuel's burn up is low enough. Various methods can be used to avoid the risk of iticality.

ANISTER FABRICATION METHODS

rial fabrication of full-sized canisters has shown that both forming from rolled plate and extrusion e possible methods for fabricating copper canisters on a full scale. In the case of forming from lled plate, it was possible with available ingots and equipment to satisfy the requirements on the icrostructure in the material to an essential degree. In the case of extrusion, the results were romising although they did not fully fulfil the objective. But there are good prospects for hieving the desired grain size in the material by means of modified process parameters and ossibly cooling during extrusion.

ANISTER SEALING METHODS

o fulfil the stringent requirements on sealing of the copper canister, a method is being developed mploying electron beam welding of the copper lid. The same method is also employed to attach e bottom of the copper canister. All development efforts are currently being concentrated on this ethod. Alternative methods that have been proposed are friction welding and diffusion bonding.

Methods for non-destructive testing of the weld are being developed in parallel in order to erify that it complies with the stipulated requirements.

Results obtained show that electron beam welding is a feasible method for fabricating and ealing copper canisters. Fully satisfactory results have not yet been obtained from the work on evelopment of methods for non-destructive testing; this will require further efforts.

The electron beam welding technique has been under development for several years at The Velding Institute (TWI) in Cambridge. The work to date has concentrated on developing a orizontal welding technique.

A reliable horizontal welding technique has been developed in the laboratory. The required olerances for the fit of the canister lid to the canister to ensure reliable weld quality requires very precise fitting of the lid on the canister. Further work is currently ongoing at TWI to develop an inclined welding technique which will greatly simplify the mechanical handling requirements for the canister lid.

The use of electron beam welding presents several challenges for the design of an active plant. Electron beam welding is done under vacuum. The main challenge for the design of the Encapsulation plant is to develop a reliable method of providing a vacuum around the disposal canister.

DESIGN OF THE ENCAPSULATION EQUIPMENT

BNFL Engineering Ltd have considerable experience in the design of mechanical handling plants for fuel and active waste. The use of electron beam welding on a canister of the proposed size and weight presents several novel design features for remote handling. These are:

a) Inerting of the canister.
To ensure that there is no corrosion within the canister for the design life of the repository it is necessary to replace the air within the canister with an inert atmosphere of Argon. This requires the development of a sealing system within the canister to retain the argon whilst the canister is within the welding chamber vacuum. The proposed method is to seal the canister insert with a bolted lid and an 'O' ring seal. The seal will have a limited life but is only required to retain the argon atmosphere until welding of the outer copper canister is complete.

b) Remotely installing the canister within the welding station vacuum chamber.
The canister is a large object, approximately 1m dia x 5m high weighing 25 Tonnes. This requires the development a transport system capable of moving the canister between process stations, remotely lifting and lowering the canister and of interfacing with the welding station to form a vacuum system. The proposed solution is to move the canister between stations utilising a shielded frame. The inner part of the shielded frame provides the lower half of the vacuum chamber. A jacking system on the frame lifts the canister into the station.

c) Sealing of the vacuum chamber.
The welding station vacuum chamber is in two parts. The lower section is part of the shielded frame, the upper part is part of the welding station. A reliable sealing system that can be remotely engaged is required. The proposed solution is to use an inflatable seal. This has the advantage that there is no hard contact between the two parts of the vacuum chamber and there is a large tolerance for misalignment between the two sections.

d) Electron beam gun.
The electron beam requires a precise beam geometry to achieve quality welds. The electron beam system must be reliable to ensure that failed canisters are minimised. Remotely positioning a large canister to the required accuracy is difficult. The proposed solution is to provide a gun with the relevant motions and sensors for self alignment with the weld and to

compensate for eccentricity and thermal effects during rotation of the canister. The design of the electron gun and vacuum system has been developed at TWI. The trials to date indicate that the required reliability and accuracy can be achieved.

) Fitting of the canister lid.
The canister lid has to be fitted remotely after charging of the canister with fuel and the inerting of the canister. The copper in the electron beam is very fluid. This necessitates having clearances of less than a millimetre between the canister and lid. If horizontal welding is used the machine tolerances between the lid and the canister are less than the practical positioning accuracy of the canister in the station and the thermal expansion of the canister due to self heating effects of the fuel. The proposed solution is to provide a heating system within the vacuum chamber to match the temperature of the lid to that of the canister and to provide a robot arm within the chamber capable of matching the lid to the canister with the required accuracy.

The possible use of an inclined welding technique would make the lid self centring within the canister. This would relieve the required accuracy for positioning of the lid. Further development s ongoing at TWI to develop this alternative weld geometry.

ENCAPSULATION PLANT AND PROCESS

The encapsulation plant is planned to be built directly adjacent to the CLAB interim storage facility. In designing the encapsulation plant, a great deal of emphasis will be placed on radiation protection for the personnel and the environment. This means, among other things, that the actual encapsulation procedure will be performed by remote control in heavily radiation-shielded compartments, called hot cells. A large part of the handling of canisters will also be done by remote control. Experience from CLAB and SFR, as well as from various foreign facilities, will be drawn upon.

The encapsulation process is being designed and engineered to deliver well-fabricated and carefully inspected canisters with fuel to the deep repository. In designing the process, special attention will be given to matters related to industrial and radiological safety. The work of designing a suitable process for encapsulation of the fuel can be divided into functional parts where different technical solutions are considered. The work of designing the plant is in progress and descriptions of the encapsulation process and the layout of the plant have been produced. SKB has employed BNFL Engineering Ltd and ABB Atom AB for this work. The following general description of the encapsulation process gives the status of the design work. The encapsulation process is illustrated in Figure 2.

The fuel is transported in storage canisters from the storage section of CLAB via the existing fuel hoist to a new pool block in the encapsulation plant. Identification of the fuel as well as measurements, and presumably some form of sorting, will be carried out in the handling pool. Water serves as a coolant and radiation shield in this step.

In the next step the fuel is taken from the water to the handling cell, where it is dried and placed in a disposal canister. In this part of the plant, where the fuel is handled freely, special requirements must be met to prevent the escape of radioactivity. The compartment is built with radiation-shielding walls and special requirements on containment and ventilation. This type of compartment is usually called a "hot cell". A type of handling machine that incorporates proven technology and meets stringent requirements on reliability and safety is chosen for handling in the cell. Special attention is given to accessibility for service and maintenance.

Transport of filled canisters is planned to take place in a channel that connects under the handling cell and the various work stations. Alternative transport systems are being studied and the choice will be made based on stringent requirements on reliability and safety. The canister is sealed during transport so that radioactivity cannot escape from the fuel.

Three work stations are planned for sealing of the canister. The first work station will contain the functions required for the exchange of air to inert gas in the canister and, if necessary, packing around the fuel assemblies. Finally, the steel lid is fitted and sealed. Thereafter there is no risk of radioactivity escaping from the fuel. This work station is also designed to meet the requirements made on a hot cell. The copper lid is placed on and sealed in a welding station. The design of the welding station will be based on the current work of developing an electron beam welding method. The material in the copper lid and the copper cylinder are melted together by an electron beam in a vacuum chamber. The last of the three stations will house equipment for inspection of the lid weld and for machining of the weld area on the canister and for removal of an improperly welded lid. The result of the sealing operation will be inspected by means of non-destructive testing using a technology that is being developed at the same time as the sealing method.

If inspection of the seal reveals that the weld is not approved, the canister is taken back to the welding station, where it is re-welded. In the event re-welding cannot be done or is not successful, the canister goes back into the process for removal of the lid and extraction of the fuel, which is placed in a new canister.

A routine check of surface contamination is planned, and decontamination of the outside of the canister will be possible in a special work station. Then the canisters are placed in a buffer store from which they can be delivered to the deep repository at a suitable pace. Handling in the buffer store is planned to be done with a radiation-shielding handling bell. The buffer store will be connected to a docking station for transport casks.

CANISTER LABORATORY

The design of the FEP has reached a stage where equipment proposals can be translated into operational plant. It is now planned to construct a Laboratory for Encapsulation Techniques (Canister Laboratory) where key process stages will be demonstrated at full size.

© IMechE 1996 C51?

A building in Oskarshamn has been purchased and is currently being prepared for installation of equipment. Initially this will comprise the Welding Station and the Weld Inspection Station and Machining Station, and design of these items for manufacture is well advanced. The majority of items that will be installed within the Canister Laboratory are being designed such that they can be re-used in the FEP.

The aim is to verify the suitability and reliability of the welding, weld inspection and canister handling equipment early in the design phase of the FEP project. The Canister Laboratory will demonstrate the ability to produce a disposal canister to the required specification, providing early confirmation of the robustness of the welding and validation systems to the regulators. The Canister Laboratory will also enable technical areas which might threaten the FEP programme and cost, to be identified and resolved early. The laboratory could also provide an inactive training facility for FEP operations personnel.

At a later date the Canister Laboratory may be expanded to incorporate elements of the Buffer Store and Handling Cell equipment. It is also planned to open parts of the facility to the public to provide more information about the Swedish Fuel Disposal proposals.

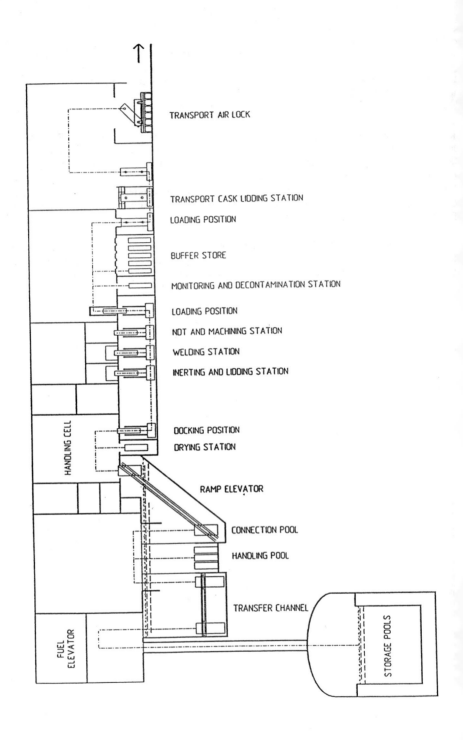

TRANSPORT AIR LOCK

TRANSPORT CASK LIDDING STATION

LOADING POSITION

BUFFER STORE

MONITORING AND DECONTAMINATION STATION

LOADING POSITION

NDT AND MACHINING STATION

WELDING STATION

INERTING AND LIDDING STATION

DOCKING POSITION

DRYING STATION

RAMP ELEVATOR

CONNECTION POOL

HANDLING POOL

TRANSFER CHANNEL

HANDLING CELL

FUEL ELEVATOR

STORAGE POOLS

Figure 2 - Encapsulation Process Diagram

Evaluation on heat removal performance of Sumitomo vault storage system

K ASHIWAGI, Y SASAKI, and T FUTAMI
Sumitomo Metal Mining Company Limited, Tokyo, Japan

Synopsis :
SMM (Sumitomo Metal Mining) is now in the stage of research and development of the MVDS (Modular Vault Dry Store) concept application to spent fuel storage here in Japan. In this paper, we report the results of a heat removal evaluation which establish the concept of a vault storage facility, satisfying temperature safety criteria and having excellent heat removal properties. First of all, it shows the analysis method employed to calculate the maximum fuel cladding temperature, and next, it makes clear the temperature distribution of every storage tube by explaining the heat transfer paths in a storage tube. Finally, it shows the results of heat removal performance with regard to the variation of stack height and storage array pitch.

Notation :

Cy = cylinder spacing normal to upstream uniform flow direction
D = tube diameter
F = air flow rate
h = average heat transfer coefficient
Nu = average Nusselt number , except the 1st and 2nd row ($=h \cdot D / \lambda$)
Re = Reynolds number ($= U_t \cdot D / \nu$)
T_f = maximum fuel cladding temperature
T_o = outlet air temperature
U_t = average velocity through between tubes ($=U_\infty \cdot Cy / (Cy-D)$)
U_∞ = average superficial flow velocity upstream
λ = thermal conductivity
ν = kinematic viscosity

1. Introduction

According to the estimation of the Science and Technology Agency(STA), the annual output of spent fuel in Japan is increasing and by the year 2005, the total amount will rise to 1.6 times that of 1994, and the total accumulation 2.1 times. It is anticipated that, in the near future, need for further storage and control of spent fuel will increase as the quantity of spent fuel arisings exceed the quantity that is set aside for reprocessing [1]. Construction of spent fuel storage facilities will be required in the future due to the limitation of storage capability of spent fuel at storage pools within reactor sites.
SMM(Sumitomo Metal Mining) is now in the stage of research and development of the MVDS (Modular Vault Dry Store) concept [2] application to spent fuel storage (PWR, BWR fuel) in Japan [3]. For the safe storage of spent fuel at vault storage facilities, the building construction must address heat removal, critical, shielding and earthquake resistant aspects of the design. The most important feature in establishing a vault storage facility is its heat removal capacity.
In the heat removal design, it must be demonstrated that the storage building concrete temperature will not exceed the guaranteed structure strength temperature. Also, that the building structure must promote the heat removal process so as to maintain the spent fuel cladding temperature below the creep critical temperature. In this paper, we report the results of heat removal evaluation which establish the concept of a vault storage facility, satisfying temperature safety criteria and having excellent heat removal properties.

2. Outline of SMM Vault Storage Facility

Figure 1 shows the concept of SMM vault storage facility. The storage building made of concrete guarantees shielding performance, which is further enhanced by locating the storage module underground. The cooling method of the vault storage facility is by natural convection within the storage module generated by a driving force which results from the decay heat of spent fuel. Thus, the decay heat of the spent fuel will be removed by convection heat transfer.

Figure 2 shows a storage tube structure. Spent fuel is packed into the canister and the cavity is filled with helium gas. Canisters are located inside a staggered array of storage tubes within the module and the air in the interspace between the canister and its storage tube is maintained at negative pressure. The air cooling flow within the storage module is horizontal to the ground and perpendicular to storage tubes.

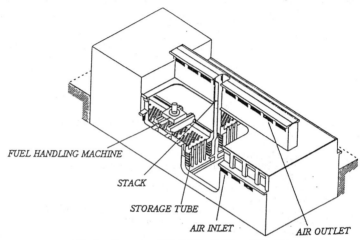

FUEL HANDLING MACHINE

STACK

STORAGE TUBE

AIR INLET   AIR OUTLET

Fig 1   Concept of SMM Vault Storage Facility

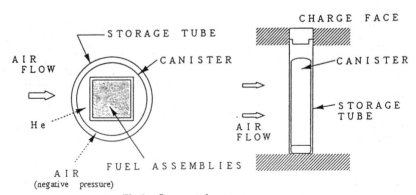

Fig 2   Storage tube structure

3. Analysis Method of Heat Removal Evaluation

Vault storage facility structure and fuel specifications used in the heat removal analysis, are shown in table 1 and 2 , respectively.

Table-1 Vault storage facility structure

Dimensions of storage module	10m(W) × 15m(D) × 6m(H)
Number of storage tubes	196 / module
Number of rows of storage tube	17 rows
Pitch of storage tubes	1.6 D
Fuel assemblies in a storage tube	1 fuel assembly
outlet stack height	30 m

Table-2 Fuel specifications [4]

Type of fuel	PWR
Number of pins per fuel assembly	17 × 17
Burn up	35000 MWD/MTU
Cooling period	5 years
Heat generation	1000 W/assembly

The procedure of the heat removal analysis is as follows. First of all, we determine the air flow rate and outlet air temperature from the balance of driving force caused by the decay heat from the spent fuel and the pressure drop of the air flow path within the storage facility.

Next, we calculate the average heat transfer coefficient around the storage tubes using an empirical formula which predicts the average Nusselt number (Nu) within a staggered tube bank [5]. This empirical formula gives average heat transfer coefficient beyond the 3rd row, and is expressed as follows. The thermophysical properties are estimated at the film temperature.

$$Nu = 0.243 \, Re^{0.64} \qquad (1)$$

$$h = Nu \cdot \lambda / D \qquad (2)$$

We have conducted a thermal conductivity and radiation analysis of the storage tube internals by applying a heat transfer coefficient obtained from the above equations as a boundary condition. Figure 3 shows this analysis model which simulates the internal features of the storage tube with a two dimensional cross section. The fuel assembly internals are regarded as uniform conducting solid by utilizing a porous medium model [6]. Conduction and radiation heat transfer have been utilized to conduct the heat transfer analysis between the fuel assembly and canister, canister and storage tube. This analysis was performed by combining the heat and fluid flow analysis code "CFDS-FLOW3D" with the radiation heat transfer analysis code "RAD 3D" released by AEA Technology.

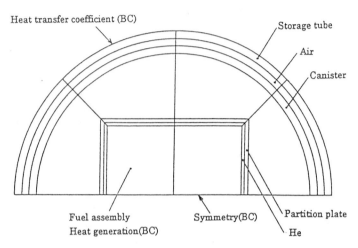

Heat transfer coefficient (BC)

Storage tube

Air

Canister

Fuel assembly
Heat generation(BC)

Symmetry(BC)

Partition plate

He

Fig. 3   Analysis model in the storage tube

4. Storage Tube Temperature Evaluation

The air cooling flow within the vault storage module is perpendicular to the storage tubes. It is known that the local heat transfer coefficient of tube banks arranged perpendicular to air flow direction has some variation in the direction of circumference and becomes almost uniform beyond the 3rd row, and that the distribution of local heat transfer coefficient changes with the tube arrangement. We have expected that the circumferential variation of local heat transfer on the surface of the storage tubes results in an uneven temperature distribution of the storage tube surface in some degree. Therefore, we have decided to examine the effect of the variation of local heat transfer coefficient on the fuel cladding temperature.

Figure 4 shows the results of the analysis giving the circumferential surface temperature distribution of the 1st row storage tube using local heat transfer coefficient distribution and average heat transfer coefficient boundary conditions. As an extreme example , this illustration also shows the result assuming the storage tube surface as a constant heat flux. The local heat transfer distribution is based on empirical results for tube banks with the same flow conditions (Re number = 30000). The temperature distribution of storage tube surface becomes large when the storage tube surface is modeled as constant heat flux. But there is almost no difference between the temperature distributions of storage tube internals in the cases of the local heat transfer coefficient around the storage tube surface and the average heat transfer coefficient. According to this analysis, the maximum fuel cladding temperature showed the same value (170 ℃) regardless of boundary conditions.

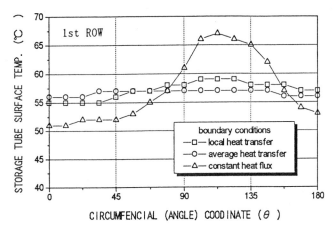

Fig 4　Temperature distributions of storage tube surface

Figure 5 shows the circumferential variation of storage tube and fuel assembly temperatures
the configuration when the local heat transfer coefficient is used for the 1st row of the storage
~es.

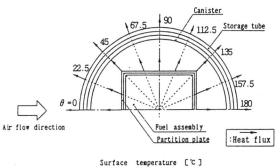

Surface temperature [℃]

θ	Storage tube a-a'	Fuel assembly b-b'
0	55	121
22.5	55	118
45	56	116
67.5	57	118
90	58	119
112.5	59	118
135	58	116
157.5	58	118
180	57	120

Fig 5　Heat transfer configuration

As the thermal conductivity of the storage tube material is large, the decay heat from th spent fuel will be removed to a larger amount at those portions where the heat transfer coefficie is high in the direction of circumference. It consequently makes the storage tube circumferenti temperature uniform. The outside surface temperature of the fuel assembly and the maximum fu cladding temperature are both almost constant regardless of the heat transfer coefficie distribution around the storage tube surface. Therefore the maximum fuel cladding temperatur evaluation can be made by using the average heat transfer coefficient without the necessity c setting the local heat transfer coefficient to the storage tube surface as a boundary condition, in th case of the internal storage tube heat transfer analysis. Moreover, similar results are obtained fc the 2nd row of storage tubes and beyond. The average heat transfer coefficient for those storag tube surfaces beyond the 3rd row is almost constant. But the storage tube surface temperature wi increase row by rows for the reason of that the local ambient air temperature around the storag tubes increases with the increase of rows.

Figure 6 shows the fuel cladding temperature for each storage tube row. Maximum fu cladding temperatures up to the 3rd row decrease and the maximum fuel cladding temperatur beyond the 4th row increases. The former is caused by the increase of heat transfer coefficient du to the development of turbulence and the latter is caused by the increase of ambient air temperatur around the storage tube. In the analysis of this specific design the fuel cladding temperature of th 15th row is higher than that of the 1st row, and the number of storage tube rows is limited by th maximum fuel cladding temperature which occurs in the last row.

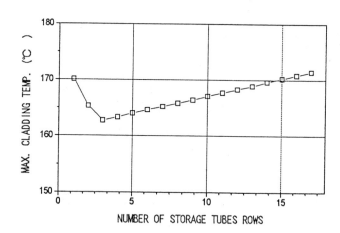

NUMBER OF STORAGE TUBES ROWS

Fig 6   Fuel cladding temperature in each storage tube row

5. Analysis for Optimum Heat Removal

We have evaluated the heat removal performance with regard to the variation of stack height and storage tube array pitch, utilizing the analysis methodology for heat removal explained earlier in this paper.

Figure 7 shows the variation in the air flow rate (F), outlet air temperature (T_o), average Nusselt number (Nu) and fuel cladding maximum temperature (T_f) of vault storage facility with regard to stack height variation. Parameters used for stack height are 10 - 50 metres from ground level. The maximum fuel cladding temperature shows in the illustration is for the last tube row, and Average Nusselt number show the value that exists beyond the 3rd row. The illustration shows the ratio of each item normalized to its maximum value for stack height variation of 10 - 50 metres. The amounts of variation are as follows.

air flow rate (F)   :   17.5 - 28.4   [m3/sec]   (F_{max}=28.4)
outlet air temp. (T_o)   :   38.2 - 43.3   [°C]   (T_{omax}=43.3)
average Nusselt number (Nu) :   115 - 166   [—]   (Nu_{max}=166)
max. fuel cladding temp. (T_f) :   168 - 179 [°C]   (T_{fmax}=179)

Figure 7 presents the air flow rate increase with increase of stack height due to the increase of driving force. An air flow rate increase of 40% and an increase of 30% in Average Nusselt number can be expected by changing the stack height from 10 metres to 50 metres. However, a decrease of only 5 °C at the air outlet temperature and 11 °C in the fuel cladding temperature is obtained in spite of the large increase in the air flow rate.

Figure 8 shows the analysis results of storage tube array pitch variation. In this analysis, the storage tube array pitch has been varied between 1.2D and 1.6D. Therefore, the number of storage tubes in the vault has varied from 196 for an array pitch of 1.6D to 357 for an array pitch of 1.2D. This illustration also shows the ratio of each item normalized to its maximum value, similar to the previous illustration. The amounts of variation are as follows.

air flow rate (F)   :   19.5 - 25.0   [m3/sec]   (F_{max}=25.0)
outlet air temp. (T_o)   :   39.6 - 51.6   [°C]   (T_{omax}=51.6)
average Nusselt number (Nu) :   147 - 227   [—]   (Nu_{max}=227)
max. fuel cladding temp. (T_f) :   171 - 177 [°C]   (T_{fmax}=177)

The pressure drop coefficient within the storage module decreases with increasing storage tube pitch, but the total heat generation decreases and results in an overall reduction in the driving force. A substantial increase in the air flow rate is observed due to the this effect up to a storage tube array pitch of 1.3D but subsides for further pitch increase. Moreover, The outlet air temperature drops due to the decrease in total heat generation and Average Nusselt number on the storage tube surface reduces due to the drop of the air velocity caused by the larger spacing between the storage tubes. Therefore, the max. fuel cladding temperature remains almost constant regardless of the variation in storage tube array pitch.

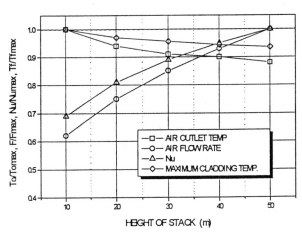

Fig. 7 Evaluation for stack height variation

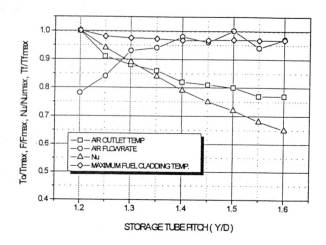

Fig. 8 Evaluation for storage tube array pitch variation

6. Conclusion

We have evaluated the heat removal performance by reference to a specific example of a vault storage facility having a PWR fuel assembly enclosed in a canister. From this evaluation, we have found the following.

- There is almost no difference between the maximum fuel cladding temperature in the cases of the local heat transfer coefficient around the storage tube surface and that of the average heat transfer coefficient.
- It is necessary to select a stack height considering both the scenery of the location and the economics, based on the fact that a significant reduction of fuel cladding temperature cannot be expected for stack height exceeding 10 metres although the heat removal performance increases with stack height.
- A more economical structure can be achieved by arranging the storage tube pitch as small as possible, should there be no problem in constructing an earthquake-proof structure.

The design conditions used in this evaluation result in the maximum fuel cladding temperature having a margin against the limiting temperature in the future. A more economical structure will be possible by increasing the number of rows of storage tubes within the storage module and the number of fuel assemblies within the canister according to the customer's needs , provided the fuel cladding and concrete temperature limits can be satisfied. The current vault storage facility has flexibility in its design.

Reference

1) STA Nuclear Office, Nuclear pocket book, 1995 192
2) M.Blackbourn,An open loop air thermosyphon cooling system for irradiated nuclear fuel in a modular vault dry storage(MVDS), ASME FED 1987 61 11-17
3) H.Nakahira , Conceptual design of vault dry storage facility, Low and Intermediate Level Radioactive Waste Management, 1993   vol 1
4) T.Saegusa, Cask Storage of Spent Fuel at Reactors, CRIEPI port, 1993 7
5) Shinya Aiba, Heat transfer around tubes in staggered tube banks, Bulletin of the JSME,1982 vol 25 No204
6) Eucken A, VDI Forschungsheft, B3, 1932 3 53

..t removal characteristics of dry storage facilities for ..nt fuel – tests with reduced scale models

AMOTO MJSocChemEng, H YAMAKAWA MJSocAtomEng, MJSocCivilEng, Y GOMI, T KOGA,
..HATTORI
.. Research Institute of Electric power Industry (CRIEPI), Chiba, Japan

Synopsis : To establish large capacity dry storage technologies for spent fuel, heat removal tests of Cask, Silo and Vault Storage Systems by passive cooling with reduced scale models have been conducted in CRIEPI. Air flow patterns in the storage area were investigated and correlations of heat transfer from a container model to cooling air were evaluated. Furthermore, parametric tests were carried out and effects of parameters on heat removal properties were investigated.

Nomenclature :

u	= air flow velocity (m/sec)	subscript m = experimental apparatus
ΔT	= rise in temperature (K)	subscript p = real facility
L	= length (m)	subscript n = natural convection
H	= length of heating zone (m)	subscript f = forced convection
Q	= heat generation (kW)	subscript z = height direction
ϕ	= degree from a front stagnant point ($^{\circ}$)	
Nu	= Nusselt number (-)	Re = Reynolds number (-)
Gr^{*}	= modified Grashof number (-)	Gr = Grashof number (-)
Pr	= Prandtl number (-)	

1. INTRODUCTION

In Japan, the importance of spent fuel storage is emphasized more than before in 'Long-Term Program for Research, Development and Utilization of Nuclear Energy' by Atomic Energy Commission, Japan, June 1994, which says 'the quantities of it (spent fuel) in excess of domestic reprocessing capacity will be appropriately stored as an energy stockpile until such time as they can be reprocessed. For the time being they will, as a rule, be stored at the nuclear power plants in the way that they have been up to now, but study will also be given to future storage methods and other necessary measures on the basis of how long it seems that they will have to be stored.'.[1] Especially, dry storage technologies with a large capacity in AFR site will be prevailing because of its economical advantage[2], and establishment of the technologies utilizing passive cooling is needed.

For the establishment, it is necessary to ascertain and evaluate the passive cooling capacity by solving issues of heat removal, and to reflect all the results into the optimum heat removal design of storage facilities.[3]

In CRIEPI, tests and numerical analyses of cask, silo and vault storage systems have been conducted for the purpose of establishing the optimum heat removal design methods and contributing the results to realization of dry storage technologies. *

2. STUDY PROGRAM

The present study program is shown in Fig 1. Tests are divided into 'Natural Convection Cooling Tests' and 'Heat Removal Tests' for Cask, Silo and Vault storage systems, respectively.

In the former, with 1/1 or 1/2 scale models of a storage container, natural convection and turbulent mechanisms near the outside surface of the container were investigated.[4] In the latter, with 1/2 and 1/5 scale models of a storage facility, convection phenomena of air flow inside the facility have been investigated, and removal of heat from the container by passive cooling have been ascertained.

Of the study program, this paper describes the 'Heat Removal Tests', whose purpose is to understand the flow pattern of air in the storage facility model and heat transfer from the cask models to cooling air.

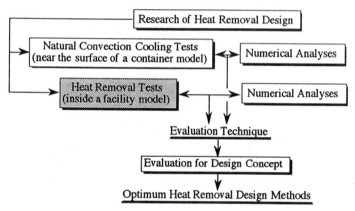

Fig 1 Study program

3. HEAT REMOVAL TEST ON CASK STORAGE SYSTEM

3.1 Experimental Apparatus

The apparatus is a 1/5 scale model of an assumed full-scale cask storage facility, in which 32 cask models are placed.

The bird's-eye view of the apparatus is shown in Fig 2. The storage facility model consists of air inlet louvers(angle:0-70degrees), storage area, and exhaust louvers. Heat-insulating materials are put on all the walls (including the floor and ceiling) of the storage facility model to prevent heat loss to the environment. The cask model is made of stainless steel which is polished as mirror surface to prevent radiation, and filled with heat-insulating materials. Rubber heaters are pasted on the inside without gap, thus a constant heat flux condition is provided. The main specifications of the cask models are shown in Table 1.

----

* This program is sponsored by Science and Technology Agency (STA) of Japan.

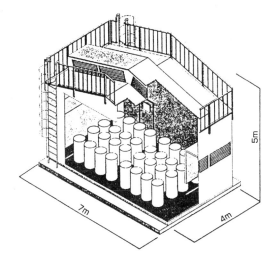

Fig 2 The bird's-eye view of experimental apparatus of Cask Storage System

Table 1 Specifications of cask models

items	specifications
Dimensions	ϕ 0.5m x 1m
Heat Generation	0 - 1.2 kW
Number	32
Materials	stainless steel
Surface	polished as a mirror

In this test, Euler number similarity is assumed, because of an importance of air flow inside the storage facility model. Under the assumption, the following relations (equation(1) and (2)) between a real storage facility and the experimental apparatus are led by equations of buoyancy force, pressure drop and heat balance, with regard to the bulk air flow rate through the storage facility model and the rise in temperature of the cooling air.

$$\frac{u_m}{u_p}=\left(\frac{Q_m}{Q_p}\right)^{1/3}\left(\frac{L_m}{L_p}\right)^{-1/3} \qquad (1)$$

$$\frac{\Delta T_m}{\Delta T_p}=\left(\frac{Q_m}{Q_p}\right)^{2/3}\left(\frac{L_m}{L_p}\right)^{-5/3} \qquad (2)$$

3.2 Test Results and Discussions

3.2.1 Flow Pattern in Storage Area

An arrangement of the cask models in the storage facility model and the flow pattern around the cask models are shown in Fig 3. In this test, smoke is introduced in the storage facility model and a laser light sheet is employed to visualize the air flow in two dimensional co-ordinates.

After recording this pattern with a video tape or a laser disk with a CCD camera, a movement of smoke particles is transferred into flow vectors using a Particle Imaging Velocimetry (PIV) technique.

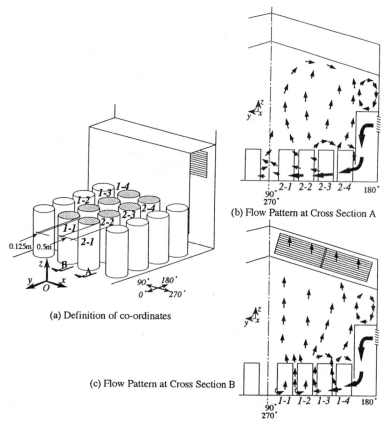

(a) Definition of co-ordinates

(b) Flow Pattern at Cross Section A

(c) Flow Pattern at Cross Section B

Fig 3 Flow pattern around the cask models

With conditions of equal heat generation for each of the 32 cask models to obtain an ideal condition, the characteristics of the flow pattern observed with above method are shown below:

· In the cross section A, the flow path in the x-direction between the cask models is wider than that in the cross section B, and horizontal air flow in the y-direction along the floor reaches to the centre of the storage area.
· On the other hand, as the flow path is narrower in the cross section B, the horizontal air flow reaches to the second cask model, at most, from the air inlet side.
· Therefore, cooling air for the third and fourth cask models in the y-direction from the air inlet side (cask model No.2-1, 2-2) is mostly supplied from the cross section A, and that for the first and second cask models (cask model No.1-3, 1-4, 2-3, 2-4) is mostly supplied from the cross section B.
· Characteristic flow patterns, depending on their positions of the cask models in the storage facility model, i.e. an up-flow along the air inlet side wall and that leaning toward

the centre of the storage area, are observed just above the cask models as shown in Fig 3. But at 2m high above the cask models, the difference of flow rate distribution is very small in the storage area, and the air vertically flows toward exhaust louvers.

Cooling air for each cask model is supplied by its buoyancy force generated on the outside surface. Thus, even if the air-supply for a cask model is momentarily interrupted and the cask model's temperature relatively rises, any steady state hot-spot does not appear because air temperature surrounding the cask model also rises and a stronger buoyancy force is generated.

3.2.2 Heat Transfer from Cask Models

For the condition of Q_{total}=17.0kW (0.53kW/cask model), heat transfer from the cask models to the cooling air is discussed below. Temperature distribution on the surface of a cask model is measured by 72 thermocouples, 8 for the circumferential distribution by 9 for the axial distribution .

(1) Local Heat Transfer Coefficient

From the flow pattern in the storage facility model as described in section 3.2.1, for a local heat transfer from a cask model to cooling air, it is necessary to consider the effect of horizontal air flow in the y-direction, based on vertical air flow (z-direction) generated by the buoyancy force on the surface of the cask model. In this test, it is assumed that the overall local heat transfer coefficient can be given by the product of local Nusselt Number for a vertical plate with a constant heat flux condition (vertical air flow by buoyancy force ; natural convection), in which the distribution of height(z) direction is considered, by a coefficient which can represent the effect of horizontal air flow (forced convection), in which the distribution of circumferential direction and y-direction are considered. That is,

$$Nu_{(z,\phi)}=C_{(Re,\phi)}Nu_{n(z)} \tag{3}$$

Here, $\cdot$ $Nu_{n(z)}$ is for a vertical plate with a constant heat flux condition (Equation for natural convection).[5]

$$Nu_{n(z)}=0.60(Gr_z^*Pr)^{0.2} \qquad (10^7<Gr_z^*<10^{12}) \tag{4}$$

$\cdot$ $C_{(Re,\phi)}$ determined from experimental data of local heat transfer coefficient is a coefficient which can represent the effect of horizontal air flow (forced convection).

$$C_{(Re,\phi)}= 0.162\left\{1- 0.26\left(\frac{\phi}{180}\right)^{0.68}\right\}Re^{0.27} \tag{5}$$

ϕ : Angle from the front stagnant points for the cask models. $0°$ is defined as follows from air flow pattern,
   a. Cask model No.1-1, 1-2 ; direction of $135°$ in the facility co-ordinates
   b. Cask model No.1-3, 1-4 ; direction of $225°$ in the facility co-ordinates
   c. Cask model No.2-1, 2-2 ; direction of $225°$ in the facility co-ordinates
   d. Cask model No.2-3, 2-4 ; direction of $135°$ in the facility co-ordinates

Re ; Reynolds number for horizontal velocity. As running to the centre of the storage area, the decrease of horizontal air flow rate is considered because of the buoyancy force of the cask models.

Comparisons of experimental data with $Nu_{(z, \phi)}$ are shown in Fig 4. At the $180°$ position on the cask model No.1-4 and 2-4, as air velocity on the air inlet side is high, Equation (3) gives a conservative evaluation. But, for other cask models, it can represent the test results well along both the axis and circumference.

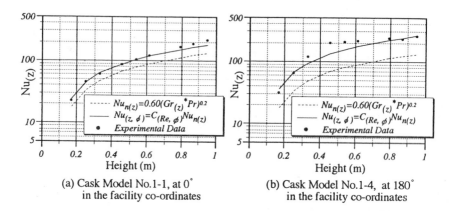

(a) Cask Model No.1-1, at $0°$
in the facility co-ordinates

(b) Cask Model No.1-4, at $180°$
in the facility co-ordinates

Fig 4   Distribution of $Nu_{(z)}$ along the axis

(2) Average Heat Transfer Coefficient

The same evaluation methodology is used for overall average heat transfer, as was used for the the local heat transfer coefficient, being given by the product of the average Nusselt Number for a vertical plate with a constant heat flux condition (vertical air flow by buoyancy force ; natural convection) by a coefficient which can represent the effect of horizontal air flow (forced convection), in which the distribution of y-direction is considered.That is,

$$\overline{Nu}=C_{(Re)}\overline{Nu_n} \tag{6}$$

Here,   · $\overline{Nu_n}$ is an average value obtained by an integration of equation(4) with height direction.

$$\overline{Nu_n}=0.75(Gr^*_{z=H}Pr)^{0.2} \tag{7}$$

· $C_{(Re)}$ is an average value obtained by an integration of equation(5) with circumferential direction.

$$C_{(Re)}=0.135Re^{0.27} \tag{8}$$

Comparisons of experimental data with $\overline{Nu}$ are shown in Table2. Although equation (6) gives a progressive evaluation of around 10 percent in maximum, it represents the test results well on the whole.

Table 2 Comparisons of average heat transfer coefficient :$\overline{Nu}$

Cask Model $N^{o\,\cdot}$	1-1	1-2	1-3	1-4	2-1	2-2	2-3	2-4
$\overline{Nu}$: Experimental Data	242	253	271	312	231	255	279	301
$C_{(Re)}\overline{Nu}_n$	249	277	297	318	245	273	294	320

3.2.3 Effects of Test Parameters

The effects of heat generation and geometrical arrangement of the cask models for heat removal performance are described below.

(1) Effects of Heat Generation

To investigate the effects of heat generation on the heat removal performance, four heat generation rate from 10kW(0.3kW/cask model) to 35kW (1.1kW/ cask model) were tested.

Results of the cooling air flow rate into the storage facility model and the rise in temperature of the cooling air at outlet are shown in Fig 5. They almost linearly increase as heat generation increases.

Furthermore, the temperature on the surface of a cask model also linearly increases as its heat generation is increased.

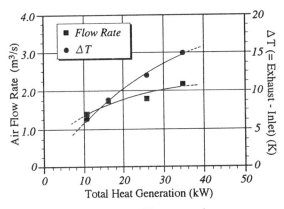

Fig 5 Effects of heat generation

(2) Effects of Geometrical Arrangement

Geometrical arrangement of casks in a real facility may be determined taking account of the operational conditions such as cask transfer. Here, effects of arrangements of cask models for heat removal performance are discussed. Three types of arrangement are considered in this test as follows :

    a. In-line arrangement with irregular spacing
    b. In-line arrangement with uniform spacing
    c. Staggered arrangement

A rise in temperature and flow rate of the cooling air for each arrangement are shown in Table3, and they were not effected by the arrangements of the cask models.

Table 3   Effects of arrangements

arrangement	heat generation (kW/cask model)	angle of air inlet louvers	ΔT (K)	air flow rate (m³/s)
In-line with irregular spacing	0.53	30°	8.6	1.83
In-line with uniform spacing	0.53	30°	8.7	1.85
staggered	0.53	30°	8.7	1.87

In-line (irregular spacing)   In-line (uniform spacing)   staggered

As a heat generation is constant, the air flow rate is determined by the pressure drop of the storage facility model. In case of 30 degrees for the angle of both air inlet and exhaust louvers, the proportion of the pressure drop produced by the in-line arrangement of the cask models with irregular spacing accounts for about 20 percents of overall pressure drop, at most. Thus, since a fluctuation of partial pressure drop caused by the other arrangements is less, the arrangement of the cask models does not have much effect on air flow rate.

3.2.4 Extrapolation of Heat Removal Performance to Real Facility

Using equations (1) and (2), the bulk flow rate and the rise in temperature of cooling air for a real cask storage facility are discussed.

For a real facility in which 32 casks are placed with the same arrangement as Fig 2, it is assumed that heat generation is 20kW per cask, and 640kW in total. On condition of Q_p=640kW, L_m/L_p=1/5, Q_m=17.0kW(0.53kW/cask model) and the test result for the in-line arrangement with irregular spacing shown in Table 3, the bulk flow rate and a rise in temperature of cooling air for a real facility are evaluated using equations (1) and (2).

· Air flow rate at inlet : 89.67m³/s
· Average flow velocity of down-flow in the duct at inlet : 0.70m/s
· a rise in temperature of air at outlet : 6.6K

4. HEAT REMOVAL TEST ON SILO STORAGE SYSTEM

The bird's-eye view of the apparatus is shown in Fig 6. Dimensions of the canister model are 1.0m ϕ x 2.5m and the heat generation rate is 20kW.

Tests with parameters of dimensions of Silo model and cooling air flow rates were carried out. Based on the test results, it was found that average heat transfer correlation from the outside surface of the canister model to cooling air is given by the following equation of Churchill's [6].

$$Nu^{1/2}=0.60+0.387\left[\frac{Gr/Pr}{\left\{1+(0.559/Pr)^{9/16}\right\}^{16/9}}\right]^{1/6} \tag{9}$$

5. HEAT REMOVAL TEST ON VAULT STORAGE SYSTEM

The bird's-eye view of the apparatus is shown in Fig 7. Four different kinds of vault storage tests can be conducted using various combinations of the device system (Open Cycle and

Closed Cycle) and the cooling air flow system (Horizontal Flow and Vertical Flow). Tests are now being carried out. The main specifications of the vault heat removal test facility is shown in Table 4.

Table 4 Specification of vault heat removal test facility

Facility Model	
· Dimension	
All facility	*4.1m x 1.3m x 9.6m*
Storage-cell-model	*2.0m x 0.92m x 1.26m*
Canister Model	
· Dimension	*0.1m ⌀ x 1.29m*
· Number/Arrangement	*39 / Staggered*
· Heat generation	*~240 watt/Canister-model*
· Surface	*Polished as a mirror*
Heat Pipe <for Closed Cycle>	
· Type	*Fin-Tube*
· Number	*4 steps*

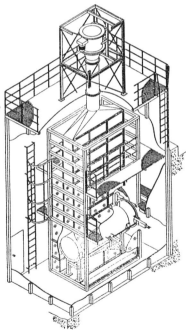

Fig 6 Experimental apparatus of Silo Storage System

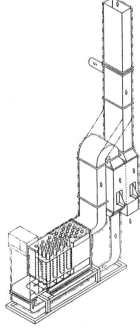

Fig 7 Experimental apparatus of Vault Storage Syste

6. CONCLUSION

Heat removal tests for the cask storage systems were carried out, and the flow pattern of air in the storage facility model and heat transfer from the cask models to cooling air were understood. Similar tests and evaluations are being made for the silo and vault storage systems. Hereafter, with these test results and numerical analyses techniques, the heat removal performance of real storage facilities will be evaluated in detail.

References

[1] Atomic Energy Commission, Japan, Long-Term Program for Research, Development and Utilization of Nuclear Energy, 1994

[2] Nagano,K., et.al., International Seminar on Spent Fuel Storage -Safety, Engineering and Environmental Aspects, Vienna, 1990

[3] Kashiwagi,E., et.al., Evaluation of Heat Removal System by Natural Convection in Spent Fuel Dry Storage Facilities, Safety and Engineering Aspects of Spent Fuel Storage, Vienna, 1994, p.418, IAEA and NEA

[4] Hattori,Y., et.al., Experiments of Natural Convection to Evaluate Heat Transfer in the spent Fuel Dry Storage Facilities, ICONE-3, Kyoto, 1994, p.1927, JSME/ASME

[5] Vliet,G.C. and Liu,C.K., An Experimental Study of Turbulent Natural Convection Boundary Layers, J. Heat Transfer, Nov.1969, vol.91, p.517

[6] Churchill,S.W. and H.H.S.Chu, Correlating Equations for Laminar and Turbulent Free Convection From a Horizontal Cylinder, Int.J.Heat Mass Transfer, 1975, vol.18, p.1049

ximum allowable temperatures of WWER-1000 spent fuel
der dry storage conditions

ADARMETOV PhD, Y K BIBILASHVILI, A V MEDVEDEV PhD, and F F SOKOLOV PhD
ochvar All-Russia Research Institute of Inorganic Materials, Moscow, Russia

ABSTRACT

In the work presented prominence is given to the problem of fuel cladding integrity during dry storage in an inert gas, as a primary barrier between the environment and radioactive fission products.

Post-irradiation thermal creep of Zr-1%Nb cladding is studied experimentally. A method is suggested to determine the time of safe long-term storage of spent WWER-1000 fuel that takes into account the pre-history of fuel assembly irradiation and cooling in water pools.

Preliminary assessments of the maximum allowable temperatures of the WWER-1000 spent fuel dry storage in inert gases are made.

1 INTRODUCTION

Nuclear power in Russia is based on three types of water cooled reactors WWER-440,-1000, and RBMK-1000. The fuel strategy of nuclear power plants that assumed the closed fuel cycle for all types of reactors has recently undergone significant changes. As before, only WWER-440 spent fuel is subject to reprocessing. Based on economics the decision was taken not to reprocess RBMK spent fuel. Consideration is given to the feasibility of interim and long-term dry storage of WWER-1000 spent fuel as cooled in water filled pools for not less than three years. For demonstration, spent fuel was stored in a gas filled cask "Kastor" and in a domestic dual purpose (storage/transportation) cask TK-13 for one year.

A method has been suggested (1) to be used for assessment of the maximum permissible temperature of spent fuel under normal dry storage conditions. The following criteria were assumed to be the criteria of safe storage:

•hoop creep strains of fuel cladding are not to exceed a 2% limit;

•elimination of through penetration of a corrosion defect from inner to outer fuel cladding surface.

The advantages of the method are the feasibility of taking into account the pre-history of in-pile fuel irradiation and its pre-cooling in a water pool.

In reference (1) the first results of studies into the post-irradiation creep of Zr-1%Nb alloy used as a cladding material for WWER-1000 fuel rods, were also described. References (2-4) indicate that irradiation induced hardening can significantly reduce the thermal creep rate in irradiated Zircaloy-2,-4. Similar results are also true for Zr-1%Nb alloy.

In this work the method is used under specific conditions of WWER-1000 spent fuel dry storage.

2 VENTILATED STORAGE CASK SYSTEM

As an example of specific dry storage technology consideration is given to Ventilated Storage Casks (VSC) system (5) of Sierra Nuclear Corporation (U.S.) offered for one of the WWER-1000 NPP.

The VSC system includes:

1. Ventilated Concrete Casks (VCC), Fig.2.1.

2. Multi-Assembly Sealed Baskets (MSB) with storage sleeves to hold 24 WWER Spent Fuel Assembles (SFA).

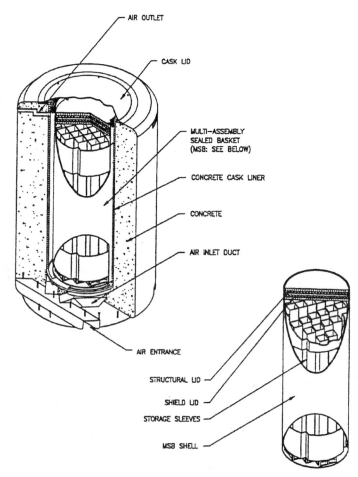

AIR OUTLET

CASK LID

MULTI-ASSEMBLY
SEALED BASKET
(MSB: SEE BELOW)

CONCRETE CASK LINER

CONCRETE

AIR INLET DUCT

AIR ENTRANCE

STRUCTURAL LID

SHIELD LID

STORAGE SLEEVES

MSB SHELL

Fig.2.1 Ventilated Concrete Cask (VCC) and
Multi-Assembly Sealed Basket (MSB)

2.1 Normal Operational Mode

The main design parameter in the VSC system, which largely dictates the behaviour of fuel in storage, is the maximum temperature of the fuel rods.

In reference (1) a method of evaluating the maximum allowable temperature of WWER-1000 dry spent fuel storage involving the following stages is offered:

- specification of the fuel element operational regimes in the core as well as time of cooling and the initial temperature of dry storage;
- determination of residual fuel clad stresses and strains by the end of the campaign and cooling period;
- calculation of creep strains and growth of cracks in clads under storage.

On the basis of the method, calculations were carried out on WWER-1000 fuel rods from a three-year campaign with maximum burn-up 50 MW•d/kgU. It is assumed that fuel was cooled for not less than 3 years in at-reactor water-filled pools.

The preliminary assessment of the maximum allowable temperature of the WWER-1000 spent fuel under dry storage in inert gases shows that the initial temperature of fuel must be ≈ 380°C.

One of the important parameters is change of initial fuel temperature during long-term storage. The real temperature drop of 24 SFAs is defined by residual heat release (time of cooling) and specific heat removal conditions (design of a multi-assembly basket and concrete cask). Fig. 2.2 shows that drops of PWR spent fuel temperature during storage in a dry steel cask TN-24P (USA production) depends on the time of SFA cooling (4). Fig. 2.2 demostrates that the main reduction in temperature of fuel occurs over the first 7 years. For the spent fuel with a longer cooling time, temperature decreases more slowly if the same initial temperature is assumed. Therefore by choice of maximum allowable temperatures and comparison of these effects in the various projects, it is possible to take into account the factor of preliminary cooling in a pool. Similar results should be expected during WWER-1000 spent fuel dry storage.

Technical requirements for a VSC design for WWER NPP spent fuel indicate a 5-year term of preliminary cooling in an at-reactor pool and an initial dry storage temperature of 380°C. In this case it is possible to expect a considerably smaller reduction in fuel temperature in storage. For example, from data in Fig. 2.2 for times of cooling during 3 years and 1 year the difference in final temperature of storage reaches 150°C, for 3 and 5 years this difference will be of the order 60°C. For 5-year cooling in a pool, the temperature of the fuel rods will be more than 300°C in inert gas over an extended period. Moreover, experiments in Russia with long-term storage (above 10 years) of uranium-plutonium fuel pellets in helium showed that degradation of the crystalline structure and microstructure of pellets takes place with considerable increase of open porosity. This circumstance may result in additional release of fission gases accumulated during operation and available from the fuel matrix at the beginning of storage. In addition, it is necessary to consider that this circumstance will be enhanced with improvement in the economic indices of the fuel cycle due to increasing fuel burn-up.

Experimental data available now show that as fuel burn-up release of fission gases into the cladding increases, in some sections over the radius of fuel pellets (the so-called periphery rim-layer), the crystalline structure of the fuel degrades completely to an amorphous state. If 100% release of fission gases from the fuel matrix into the fuel rod cladding is predicted, then the internal pressure under fuel rod cladding will be 11 MPa at 20°C. This value shall be used in calculations for conservative estimations of spent fuel temperature during long-term storage, especially for fuel with high burn-up of >55-60 MW•d/kgU.

Fig.2.3 shows the calculational relationship of maximum initial temperature of WWER-1000 spent fuel in dry storage, obtained on the basis of the criterion that hoop creep strains of fuel rod cladding material will not exceed 2% during the whole period of storage. It follows from the figure that the maximum permissible temperature of spent fuel will not exceed 330°C under long-term, >10 years, storage in the medium of inert gases.

2.2 Off-Normal Events

These modes include abnormal events, design-basis accidents and transients.

The transients are realized in operations with spent fuel during vacuum drying and filling an MSB with helium. The maximum temperatures reached are here 437°C in a vacuum and 404°C in helium. These are the most important modes, because they correspond to maximum temperatures and all SFAs pass through them before long storage. As well as in normal conditions the integrity of fuel is evaluated by comparison of calculated temperatures with the accepted maximum allowable temperature. In the project it is defined at a level of 570°C for 8 hours. It is assumed that, "within the limits of this temperature there will be no damage due to the effect of temperature".

At a level of residual SFA heat ≈1.18 KW (after 5-year cooling), during re-loading in a basket and vacuum drying of a basket when SFA cooling is provided by free convection in air environment, it is essential for the oxidation reaction that the heating of fuel rods does not occur. Thus, increased temperature of fuel rods at a transient arises only in an inert environment and will not result in additional oxidation of cladding. Under these conditions the process determining the requirements of leak-tightness of cladding, will be high-temperature creep of a cladding material due to internal pressure.

The development of deformation processes occurring in cladding due to overheating, was simulated in experiments on heating of tubular samples pre-irradiated to different flux levels and loaded by internal pressure. In experiments, samples from alloy Zr-1%Nb 9.15 mm dia 0.7 mm thick were used, loaded at normal conditions (20°C) with a pressure of argon equal to 5, 7 and 9 MPa and irradiated in a research reactor BOR-60 at temperature of ≈350°C. After irradiation a number of samples (No. 1-10) were tested in a vacuum furnace over 500 hours to evaluate the fuel rod behaviour at normal dry storage conditions (1). The results of experimental investigations into transients during dry storage of fuel rods are given in Table 2.1.

Tests have shown that the effect of flux level on post-irradiation thermal creep, and also on rupture and final deformation of samples, is not statistically significant. This is explained by a weak feature of the structure of radiation defects in Zr-1%Nb alloy, which interfere with dislocation movement during thermal creep, and the quantity of which reaches saturation as a result of irradiation to high flux levels.

The preliminary assessment of data shows that the maximum allowable temperature of transients during dry storage of WWER-1000 spent fuel in the VSC system for WWER-1000 should be reduced to 120-130°.

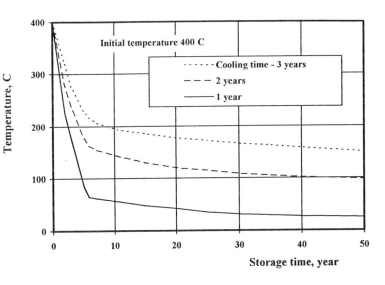

Fig.2.2 Temperature profiles for the spent fuel with cooling times of 12, 24 and 36 months

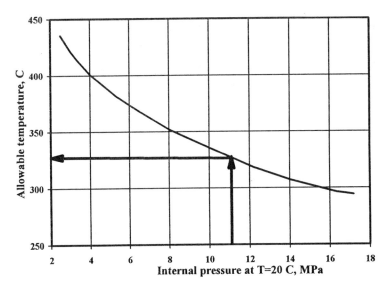

Fig.2.3 Maximum allowable temperature of WWER-1000 spent fuel during dry storage

3 CONCLUSIONS

Results of experimental investigations into thermo-mechanical properties of pre-irradiated Zr-1%Nb alloy over a range of temperatures 500-570°C are presented.

Safety examination of the VSC dry storage system has been carried out. Preliminary safety criteria under dry storage conditions in an environment of inert gas are as follows:

•maximum cladding temperature under normal conditions of dry storage should not exceed 330°C after 5-year cooling in water-filled pools;

•maximum allowable temperature of spent fuel rod cladding under operational mode with infringement of heat removal should not exceed 440°C over 8 hours.

As each SFA dry storage project comprises its individual technology of spent fuel management, it is necessary to evaluate allowable parameters (terms of storage, maximum temperatures of fuel) for each project respectively. The programme of experimental investigations for the justification of safety criteria for WWER-1000 dry spent fuel storage systems is underway.

REFERENCES

1. KADARMETOV, I.M., et.al., Evaluation of Maximum Allowable Temperature of WWER-1000 Spent Fuel under Dry Storage Conditions, Spent Fuel Storage: Safety, Engineering and Environmental Aspects (Proc.Sem. Vienna, 1994), IAEA, Vienna.

2. EINZIGER, R.E., et.al., High Temperature Post-irradiation Materials Performance of Spent Pressurised Water Reactor Fuel Rods under Dry Storage Conditions, Nucl. Tech., 1982, v.57, No.1.

3. PEEHS, M., et.al., LWR Spent Fuel Behaviour, J. Nucl. Mater.,1986, v.137, No.3.

4. MAYUZUMI, M., et.al. A Method for Evaluating Maximum Allowable Temperature of Spent Fuel during Dry Storage Condition, Spent Fuel Storage: Safety, Engineering and Environmental Aspects (Proc.Sem. Vienna, 1990), IAEA, Vienna.

5. LEE, W.J., et.al., Design and Initial Operational Experience of the Ventilated Storage Cask (VSC) System, Spent Fuel Storage: Safety, Engineering and Environmental Aspects (Proc.Sem. Vienna, 1994), IAEA, Vienna.

Table 2.1 Creep strains and durability of pre-irradiated Zr-1%Nb tubular samples

No. of sample	Gas pressure, MPa	Sample dia, мм	Flux level, 10^{22} см$^{-2}$	Temperature, °C	Hoop stresses, MPa	Testing time, min — basic	Testing time, min — additional to rupture	Strain, %
1	9.0	9.24	0.96	500	157	480	-	1.9
							850	34
2	9.0	9.405	2.20	550	170	-	55	38
3	9.0	9.57	3.00	525	168	-	150	35
4	7.0	9.235	0.96	500	122	480	-	0.5
						960	-	1.6
						1440	-	4.0
5	7.0	9.6	2.20	570	139	-	35	35
6	7.0	9.33	3.00	550	131	480	190	38
		9.36		525	128	960	-	4.4
							1080	19
								39
7	5.0	9.215	0.96	570	95	-	190	68
8	5.0	9.265	2.20	570	95	-	225	60
9	5.0	9.31	3.00	570	96	-	254	49
10	5.0	9.32	3.00	525	91	480	-	0.6
						960	-	1.4
						1440	-	3.0
						1920	-	5.7
		9.85		550		-	390	37
11	9.0	9.605	4.7	550	174	-	125	39
12	9.0	9.685	1.9	525	170	-	165	45
13	9.0	9.94	3.5	570	177	-	35	25
14	9.0	9.66	1.9	500	165	-	1000	35

nstruction of an interim spent fuel store at PAKS nuclear wer plant

DÖGH
ERÖTERV Rt, Budapest, Hungary
ABÓ
Nuclear Power Plant Limited, Paks, Hungary

Four VVER - 440 MW reactor units are operated by Paks Nuclear Power Plant Ltd in Hungary.

The spent fuel assemblies from these reactors were initially transported back to the Soviet Union and more recently to Russia for reprocessing.

In preparation for the eventuality that transportation back to Russia may no longer be possible Paks NPP began to search for alternative options.

A description of the whole process is given from making the decision on interim storage, going on to the selection of a suitable system, passing through the licensing procedure and ending with the implementation phase.

1. INTRODUCTION

Hungary with an area of 93,036 km^2 situated in Central Europe in the Carpatian Basin, borders on seven countries (Figure 1).

Following World War II, Hungary, a country having interests with and under the influence of the Soviets, began to recover the from the damage caused during the war.

With the advances in the use of the nuclear energy for civil purposes and considering the shortage of energy sources within Hungary, the country's power engineering management's attention was turned towards the atomic energy. Following this the government of the time established the National Atomic Energy Commission in 1950 with a programme for making preparations for the use of nuclear energy as part of Hungary's spectrum of generation capabilities.

In 1996, under an Intergovernmental Agreement between Hungary and the Soviet Union, the Soviet Union undertook to supply design, equipment and fresh fuel for a new nuclear power plant to be constructed in Hungary.

Under similar agreements Soviet designed nuclear power plants were also constructed in the Central and Eastern European countries (Czechoslovakia, GDR and Bulgaria) within the region of Soviet interest.

Following the selection of Paks as the site for the nuclear power plant in 1967, the construction of the first Hungarian NPP came to a standstill for a while. The construction work was resumed in 1972 and four VVER 440 units have been built on the Paks site. The first, Unit 1 was commissioned in 1982 and the last, Unit 4, in 1987.

The "standstill" mentioned above, at the early stage of construction, proved to be very useful. Instead of adopting an earlier Soviet design, Hungary constructed the advanced V 213 reactors which are considered to be of the most successful type within the VVER reactor family.

Based on the safety review carried out in the early nineties, the Paks NPP performance is comparable with that of its Western counterparts.

With regard to availability the units are among the best in the world.

The domestic importance of the Paks NPP is emphasised by the fact that since 1988 it has been generating more than 40% of Hungary's electricity.

2. INITIAL SPENT FUEL STRATEGY

According to the fuel strategy that was effective at the time of the NPP construction the Soviet Union undertook not only to supply new fuel but also accepted the return of the spent fuel for reprocessing. This arrangement can be considered as unique to the relationship between the former Soviet Union and the Central-and Eastern European countries operating VVER reactors constructed with Soviet assistance, because the final disposal of waste from reprocessing was also undertaken by the Soviet Union.

Each VVER 440 reactor contains 349 fuel assemblies. A pictorial presentation of the fuel assembly used in the VVER 440 reactor is shown in Figure 2. During the annual refuelling outages an average of 116 spent fuel assemblies are replaced with fresh ones. In line with the original fuel cycle strategy the storage capacity of the decay pools located next to the reactors can accommodate fuel assemblies for three years operation. Following an initial period of storage in the pools the spent fuel assemblies were then loaded into casks designed for the transport of VVER 440 FA's and returned in specially designed railway coaches.

This concept was slightly changed in the early eighties. Due to the shortage of reprocessing capacity the spent fuel assemblies could only be returned to the Soviet Union after a cooling down period of five years.

In order to increase the time of under water storage the soviet designers suggested that a pool be built with ten years of storage capacity. Such pools had been constructed in several other countries operating VVER reactors. To meet this requirement for increased storage time under water the pool capacity was doubled at the Paks NPP. The original spent fuel racks of Unit 1 could relatively easily be replaced by new ones providing compact storage because reactor Units 2, 3 and 4 were already being serviced by the new racks. Before replacing the racks of Unit 1, spent fuel stored in the pool of Unit 1 had to be transferred into the decay pool of Unit 2 using a special container.

The first railway train transporting spent fuel back to the Soviet Union pulled out in 1989. Following this, the spent fuel was returned in compliance with the schedule for a further two years.

As the spent fuel to be reprocessed was not originally classified as waste material the reprocessing service was provided free of charge by the Soviet Union. Later a so called compensation fee was introduced by the Soviets which was afterwards modified by the "world market price" concept. This latter term does not accurately described the above-mentioned approach of spent fuel repatriation as there is no other case known, in the history of spent fuel reprocessing, when the waste is disposed of by the country providing reprocessing service.

The political and economic changes that occurred in the Soviet Union in the early nineties could cause far greater problems to Hungarian power generation than the growing costs of transportation.

3.   INVESTIGATING ALTERNATIVE SPENT FUEL STRATEGIES

Hungary faced the first difficulties related to reprocessing in 1992 when the number of FA's returned was less than planned. The potential difficulties with the re-transportation were identified just in time by the Paks NPP experts and they started to review the spent fuel strategy looking for new alternatives in case the earlier transportation arrangements were terminated. After studying the solutions applied by countries having considerable experience in the peaceful use of nuclear energy the interim storage was selected because it also meets the requirements of the International Atomic Energy Agency.

Following this exercise, potential suppliers were identified by the NPP experts and in 1991, seven foreign company were requested to prepare feasibility studies based on standardised requirements. In addition to the NPP experts several domestic institutes and companies, the authorities concerned and IAEA experts were involved into the evaluation of the feasibility studies.

A special scoring system consisting of ten most important criteria was defined by the Paks NPP experts with the varying importance of each criterion accounted for by appropriate weighing factors.

It should be emphasised that the selection was not an easy task for several reasons.

a) Different basic concepts were offered, namely wet and dry storage.

b) Different structures were offered, namely cask and vault.

c) All offers were considered acceptable from both the point of nuclear safety and technical requirements.

d) All offers had appropriate licensing backgrounds. It was impossible to check their licensing conformance for such type of facilities since Hungary did not have any regulations controlling such projects.

The selection process for the storage facility was determined on the basis of the following three aspects:

-   expectations of the parties, especially the authorities, participating in the evaluation

-   scores given by the Paks NPP experts

-   the anticipated general changes in the back-end fuel cycle strategy.

The local authorities had several expectations e.g. the storage facility must be of robust design, air should be used as the coolant and, possibly the most important, that the FA's cladding temperature of the FA's should be low. This latter point is related to the Zr-Nb cladding of the VVER fuel assemblies which differs slightly from the Zircalloy used in Western countries. Compared with Zircalloy clad FA's there is limited R & D experience available for the VVER fuel assemblies that can be used to define the maximum allowable storage temperature. In spite of the fact, that based on its standard features, the Zr-Nb alloy seems to be more favourable than Zircalloy, it was generally accepted that the greater the difference between the storage temperature and the maximum temperature of $350^{\circ}C$, specified for VVER FA's the higher would be the score allocated to a given storage technology.

Another requirement related to the anticipated change in the fuel strategy was that the storage facility must be modular in order to provide the possibility for storage of all the spent fuel used in the Paks NPP throughout it's life.

As a result of the selection process the offer from GEC ALSTHOM ESL Ltd (UK) was considered the most favourable by PA Rt, the operator of the Paks NPP. A Contract between PA Rt and GEC Alsthom ESL Ltd for the Licensing Design, the Pre-Construction Safety Report and the Construction Design for a Modular Vault Dry type of interim spent fuel store was placed on 28 September 1992. The arrangement and operation of Paks Modular Vault Dry Store are shown on Figures 3, 4 and 6.

Due to the lack of specific requirements for licensing of the planned facility PA Rt, the applicant initiated preliminary discussions with the OAH NBF, the authority responsible for issues of nuclear safety, even before signing the contract with GECA. Based on an agreement between the authority and PA Rt the licensing documentation prepared by GEC Alsthom was to be governed by Regulatory Guide 3.48 issued by the US NRC.

During the same period Paks NPP faced the first minor delay in the transportation of the spent fuel. However still trusting in the continuing return of spent fuel to Russia it was acceptable for PA Rt that GEC Alsthom should not have to take into account specific demands (e.g. the use of Hungarian materials in compliance with Hungarian Standards) during the Construction Design phase. This meant that GEC Alsthom had only to comply with the relevant British and US standards.

PA Rt's main objective at this time was to make sure that the selected storage facility had the necessary license from the authority and that the design documentation which could be domesticated at a later date was available.

Considering the future course of VVER fresh fuel R&D activity and the possible use of Western fuel assemblies, as a result of fresh fuel supply diversification, three fuel assembly scenarios were defined by PA Rt experts at the very beginning of the selection process. The design parameters for the storage facility has to accommodate all three scenarios.

At the request of the National Atomic Energy Commission a review process of the Paks-site seismic conditions and behaviour was undertaken concurrently with the selection of a spent fuel storage design.

Research data available for the Paks site, at the time of the biding for feasibility studies and the data also used in the technical section of the contract made with GEC Alsthom ESL Ltd it was stated that an earthquake of MSK 6.4 strength could not be ruled out and that an acceleration of 0.3g had to be accommodated.

Because of disagreement between the institutes, participating in the seismic review process, the final report issued in 1993 conservatively established a horizontal acceleration for the Safety Shut-down Earthquake (SSE) at 0.35g.

4. LICENSING OF THE SPENT FUEL STORAGE

Witnessing the legislative changes that were occurring within Russia, and the suspension of spent fuel repatriation to the former Soviet Union for reprocessing from countries operating VVER reactors (except Finland), PA Rt recognised that construction of the MVDS project had to commence as soon as possible. PA Rt therefore decided to modify the contract with GEC Alsthom ESL Ltd and requested them to redesign the storage facility for the new SSE value. This change to the SSE specification caused a considerable rise in costs for PA Rt both in the area of design and construction work. As a result the licensing process, required to initiate the construction work, was started in 1993.

Several domestic companies were approached by PA Rt to review and evaluate the PCSR, submitted by GEC Alsthom, with respect to conformance with RG 3.48. Following consultation, organised with the participants mentioned earlier, a new issue of the PCSR was prepared by GECA. This material served as the base documentation for the licensing procedure.

During the licensing process numerous problems had to be solved. These can be briefly described as follows:

• Contrary to the two-stage Hungarian practice the US regulation on interim storage specifies a single-stage licensing procedure. Requirements on the content of the licensing documentation are defined within the single-stage procedure of the US NRC Regulatory Guide 3.48. The use of RG 3.48 for a two-stage licensing procedure resulted in non-conformance in some cases between the PCSR, supplied by the designer, and the requirements of the Hungarian authority.

• Based on GEC ALSTHOM ESL's English language PCSR the authority required the PCSR in Hungarian to be amended or modified where necessary. For these changes to the PCSR it seemed to be less complicated to charge ETV-ERŐTERV Rt, a Hungarian designer who had taken part in the NPP design work, with the task of modifying the Hungarian PCSR rather than supply extensive data to GEC ALSTHOM ESL.

• Changes in the Hungarian Law during the licensing process required more permits to be obtained. These had not been specified at the early stage of design.

• Basically PA Rt had to obtain permits from four different licensing authorities, they were:

  - environmental permit
  - building permit
  - water permit
  - nuclear safety permit

A number of other specialised and associated authorities were also involved in the licensing procedure by the above mentioned four authorities. The licensing procedure was extremely complicated and PA Rt had to address more than 20 different authorities. Some authorities participated not only in one but sometimes two or even three licensing procedures. The Nuclear Authority's requirements on the content of specific documentation were relatively well defined by the use of Regulatory Guide 3.48. Occasionally on the other hand, in the case of specialised and associated authorities, the content of specific documentation had to be agreed upon one by one.

Following the licensing procedure, which was delayed by detailed difficulties, the construction of the Paks MVDS started when the project implementation license was granted in February 1995 by the National Atomic Energy Commission. The licensing process flow chart is shown on Figure 5.

5. CONSTRUCTION OF THE SPENT FUEL DRY STORE

Preparatory ground work began at the same time as the start the licensing procedure started. This involved terrain correction work, installation of the public utilities, earthwork and later the soil improvement work. Soil improvement was required because we had a 10 m depth of loose silty sandy soil at the MVDS site which would be inclined to liquefaction in a Safety Shut-down Earthquake of 0.35g. To ensure the appropriate compactness of the sand 8 - 10 m long gravel piles (stone columns) have been installed on a 2.5 m square lattice using Frankie technology.

During the licensing procedure PA Rt. commissioned ETV-ERŐTERV Rt. to both prepare the Hungarian version of the PCSR, for submittal to the Nuclear Authority, and also the project's domesticated implementation design documentation, which was based on GECA's Construction Design.

Justification for this approach to preparatory work is demonstrated by the fact that one month after the issue of the construction license concreting of the reception building had begun.

The MVDS total furnished capacity permits 4950 fuel assemblies to be stored. This capacity is being based upon the quantity of spent fuel produced by the Paks reactors over a period of 10 years. The store will be constructed in 3 phases. During the first phase the Transfer Cask Reception Building (TCRB) and a 3 vault module required to store 1350 fuel assemblies will be constructed. The total furnished capacity of all 3 phases will be completed by the construction of 8 more vaults constructed at a time. Each vault is capable of accommodating 450 fuel assemblies.

If it is necessary, provision is made to extend the MVDS capacity to a total of 14850 storage positions. This capacity would permit interim storage of all spent fuel assemblies produced by the four Paks reactors over thirty years of operation (see Figure 6).

One of the main advantages of this store is that the number of vaults can be extended in a modular configuration. The new modules can be constructed as needed. The "in line" arrangement of the modules allows for the use of a single FHM[1] and central TCRB facility throughout the planned extension phases.

Construction of the store is currently to schedule with commissioning of Phase I planned for the 20th of December 1996. Conformance with this schedule shown in Figure 7 requires a well co-ordinated and controlled management of the construction and licensing activities.

Construction work and manufacturing, with the exception of the FHM and part of the Fuel Drying System, are being carried out by certified domestic companies. The FHM and some components of the FDS are being manufactured by GECA who are also involved with other project activities within the frame of a Project Assistance Contract. GECA's involvement is being co-ordinated through their resident engineer.

The specific costs of construction of each phase of MVDS are presented in Figure 8. It can be seen from this Figure that the specific costs include the costs incurred with the design, the licensing, the construction and the commissioning of each phase of MVDS.

There is still a number of areas to be addressed with respect to licensing as we are required to obtain manufacturing, import and installation permits for all of the Important to Safety (ITS) items in accordance with existing regulations. Furthermore, the MVDS commissioning can only begin upon the issue of a special license.

A decision needs to be made, on whether or not to proceed immediately with Phase 2 and 3 of the extension, before the completion of Phase 1 as the capacity of Phase 1 only allows for storing spent fuel assemblies presently located in the decay pool. This decision will be made by PA Rt senior management in the near future.

The store's planned life cycle is 50 of years. Within this period the methods for final disposal of high, medium and low activity wastes, including the spent fuel assemblies stored in the MVDS, should finally be determined.

With the involvement of the relevant Ministries, a national project has been initiated to select, prepare and implement the safest, most appropriate and economic solution. This project should guarantee that in Hungary disposal of spent fuel assemblies will be managed whilst taking into consideration international recommendations and safety.

---

[1] Fuel Handling Machine

Figure 1: Map of Europe with Hungary highlighted

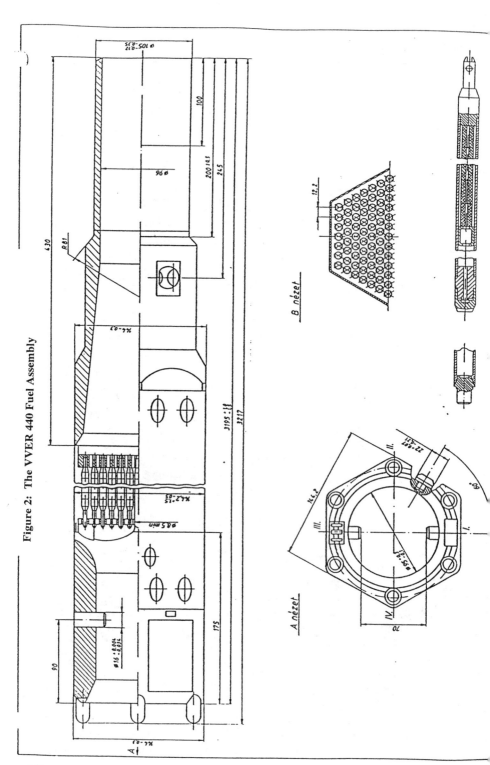

Figure 2: The VVER 440 Fuel Assembly

244

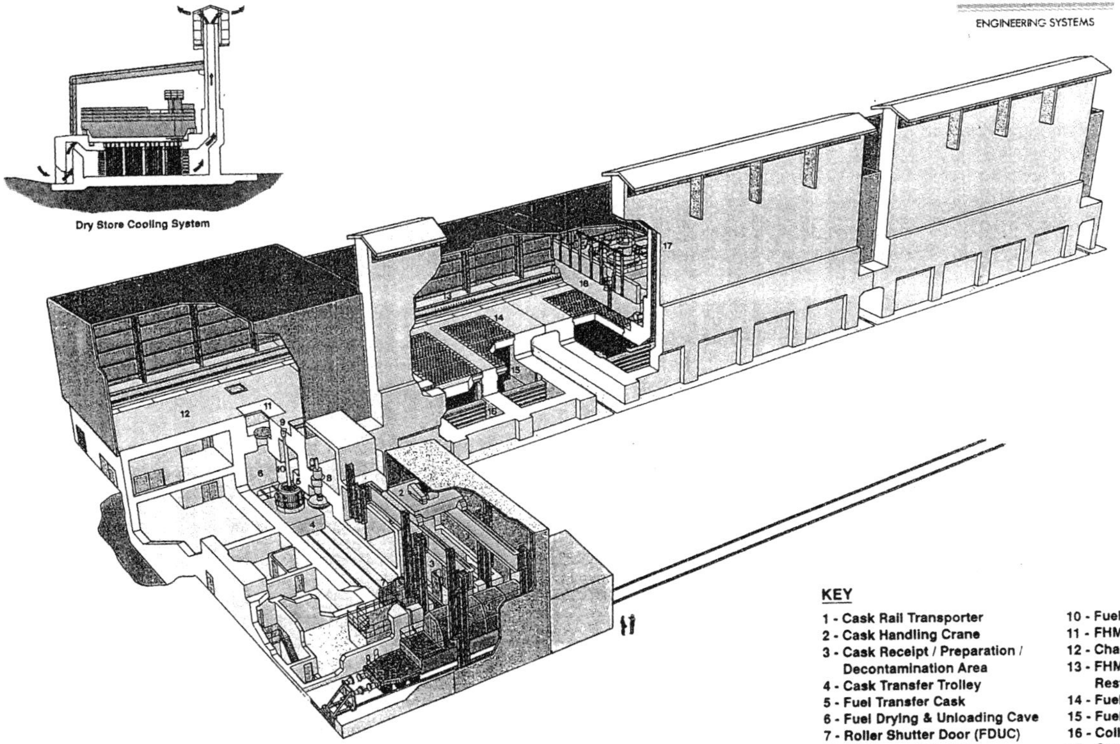

KEY

1 - Cask Rail Transporter
2 - Cask Handling Crane
3 - Cask Receipt / Preparation / Decontamination Area
4 - Cask Transfer Trolley
5 - Fuel Transfer Cask
6 - Fuel Drying & Unloading Cave
7 - Roller Shutter Door (FDUC)
8 - Lid Removal Position
9 - Load / Unload Port

10 - Fuel Drying Tube
11 - FHM Maintenance Access
12 - Charge Face
13 - FHM Long Travel and Seismic Restraint Rails
14 - Fuel Storage Vaults
15 - Fuel Storage Tubes
16 - Collimators (at Inlet & Outlet)
17 - Cooling Air Outlet Stack
18 - Fuel Handling Machine

Dry Store Cooling System

Figure 3: Paks Modular Vault Dry Store

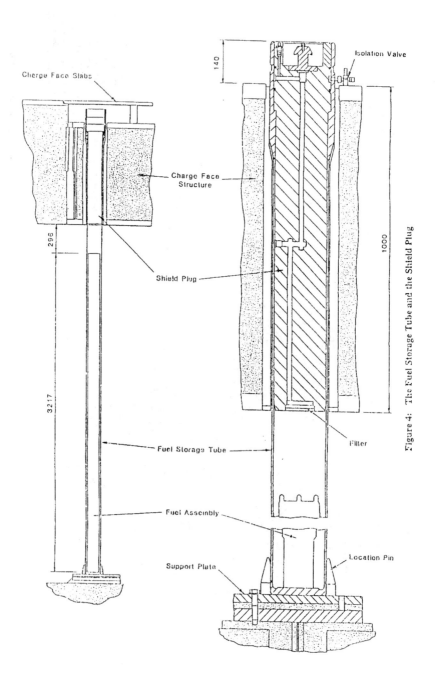

Figure 4: The Fuel Storage Tube and the Shield Plug

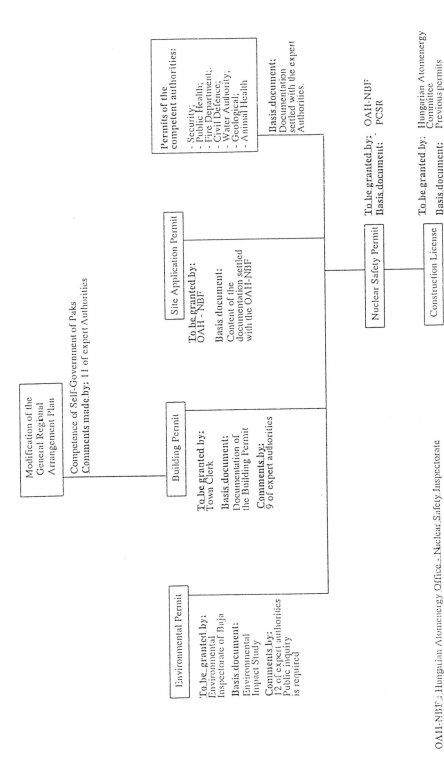

Modification of the General Regional Arrangement Plan

Competence of Self-Government of Paks
Comments made by: 11 of expert Authorities

Environmental Permit

To be granted by:
Environmental Inspectorate of Baja

Basis document:
Environmental Impact Study

Comments by:
12 of expert authorities
Public inquiry is required

Building Permit

To be granted by:
Town Clerk

Basis document:
Documentation of the Building Permit

Comments by:
9 of expert authorities

Site Application Permit

To be granted by:
OAH - NBF

Basis document:
Content of the documentation settled with the OAH-NBF

Permits of the competent authorities:

- Security;
- Public Health;
- Fire Department;
- Civil Defence;
- Water Authority;
- Geological;
- Animal Health

Basis document:
Documentation settled with the expert Authorities.

Nuclear Safety Permit

To be granted by: OAH-NBF
Basis document: PCSR

Construction License

To be granted by: Hungarian Atomenergy Committee
Basis document: Previous permits

OAH-NBF : Hungarian Atomenergy Office - Nuclear Safety Inspectorate

MODULAR VAULT DRY STORE

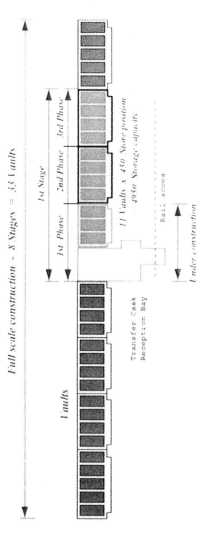

Full scale construction - 8 Stages = 33 Vaults

Vaults

1st Stage

1st Phase | 2nd Phase | 3rd Phase

11 Vaults x 450 Store position
= 4950 Storage capacity

Transfer Cask
Reception Bay

Rail access

Under construction

N

Figure 6: Cut-away th...

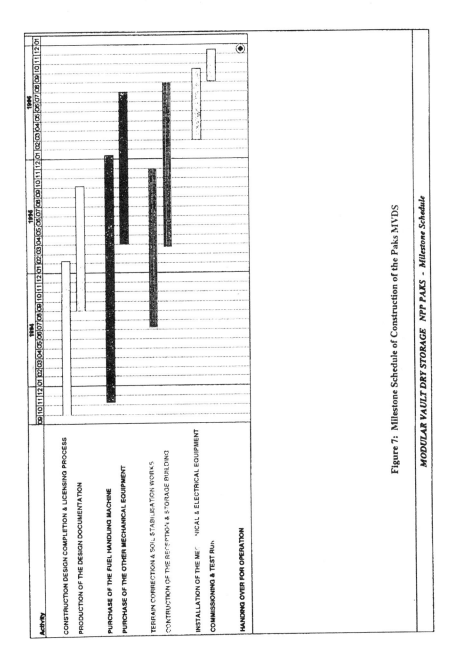

Figure 7: Milestone Schedule of Construction of the Paks MVDS

MODULAR VAULT DRY STORAGE NPP PAKS - Milestone Schedule

PAKS - MVDS SPECIFIC COSTS

USD / Kg Heavy metal

262 — I. STAGE 1 PHASE
174 — I. STAGE 1+2 PHASE
150 — I. STAGE 1+2+3 PHASE
122 — FULL SCALE IMPLEMENTATION (33 VAULT)

12/034/96

₴ regulation of storage of radioactive materials in the UK

IASON Bsc, Msc, MSRP and **B J DELANEY** Bsc, CEng, MIChemE
า and Safety Executive, HM Nuclear Installations Inspectorate, UK

SYNOPSIS

In the UK the storage of radioactive materials and waste on a licensed nuclear site is regulated and controlled by HM Nuclear Installations Inspectorate (NII) on behalf of the Health and Safety Executive. This paper considers the issues arising from the Government's Review of Radioactive Waste Management Policy and how a similar approach could be applied to some other radioactive materials such as plutonium. It describes the legislative framework, the safety standards which apply, NII's position on waste management and the role taken by the Inspectorate in the different phases in the life of a plant.

1. THE LICENSING AND REGULATORY PROCESSES IN THE UK

1.1. The Regulatory Framework

Nuclear installations are subject to legal control through "conditions" which are attached to a site licence. The law allows HSE to attach any conditions that may be necessary in the interests of safety. Three of the standard conditions specifically cover accumulation, disposal and leakage of waste and two cover nuclear matter. They establish a framework of requirements and make the licensee, as operator, responsible for the production and application of detailed safety standards and procedures and confer various powers that can be used by NII to ensure safety is properly managed.

In circumstances where NII considers that safety may be compromised, for example by the breach of a licence condition, then use can be made of other powers of the licence to exert control and secure safety, (e.g. by using a direction requiring the licensee to shut down a particular plant or process). The Inspectorate also has discretion to use the more general powers available under the Health and Safety at Work etc Act 1974 (HSWA) through the issue of Improvement Notices or Prohibition Notices.

Improvement Notices may be issued when an Inspector believes there is a contravention of a statutory provision. Failure to provide improvements within the specified timescale may result in prosecution. Prohibition Notices may be issued when an Inspector believes there is or will be a risk of serious personal injury. When issued, the activity must cease, and failure to act in accordance with the notice may result in prosecution.

1.2 The Development of the Safety Case

The licence conditions require the production of safety cases so that a licensee can satisfy itself that safety is being properly managed, that appropriate standards have been set and met, that hazard identification has been adequate, that suitable and sufficient safety features are in place, that all significant assumptions have been identified and all instructions, limits and conditions required to maintain operations within specified margins have been identified. The safety case is also one of the means by which the licensee demonstrates safety to the regulator.

A safety case is the totality of documents prepared by a licensee to assist in the consideration of safety, the preparation for, and the justification, implementation and development of each phase in the life of a site plant, including design, construction, commissioning, operation, shutdown and care and maintenance to clean-out and decommissioning. When the safety case is formed from a compilation of documents, it will always include a summary in the form of a safety report. This report and the other documents forming the safety case may make reference to supporting arguments, evidence obtained from research and development and existing standards, procedures and instructions.

The safety case should be developed by the licensee when a plant or activity moves from one phase of its life cycle to another. It should be updated or amended to take into account the activities in the next phase or any modifications to plant or procedures. In some situations supplementary documents can be used to justify a particular stage.

One licence condition requires that safety cases are subject to periodic review and assessment. The intention here, and it is particularly relevant to the long term storage of radioactive materials, is that there is a **look back** and **look forward** over operations. The **look back** should ensure that:

- there is no cumulative effect associated with modifications, and,

- lessons are learned from incidents and operations;

Furthermore, the safety of the plant or process should be compared with current engineering standards, codes, safety standards and criteria. Thus, reasonably practicable improvements can be identified and made. The strategic **look forward** should take into account:

- the projected life of the plant;

- the required life of the plant;

- the availability of waste disposal and other downstream processing routes, and,

- contingency requirements if the life of the plant is less than the period that its radioactive inventory has to be stored for.

The periodic review of safety is particularly relevant to the long term storage of radioactive materials and is consistent with NII policies in this respect.

1.3   Inspection and the Regulatory Process

The regulator will use a safety case to help judge how well safety is being managed, and to determine the extent and way in which the licensee should be regulated to allow them to carry out particular activities. Other factors are also taken into account when considering the extent of regulation, such as the novelty, complexity, scale, hazard and significance of the proposal, as well as the experience of the licensee.

The Inspectorate's most frequently used regulatory powers stem from the conditions attached to the site licence and from the arrangements made by a licensee in accordance with the requirements of the licence conditions. For example, the licence conditions dealing with the construction or installation of new plant which may affect safety requires the licensee, amongst other things to make and implement adequate arrangements to control construction or installation. Usually, the construction or installation is divided into stages, with a series of "hold points". These can be specified by NII and are normally agreed with the licensee. Progress beyond a "hold point" requires the NII's agreement.

The licensees' arrangements normally make provision for them to submit safety cases for specified stages to the Inspectorate and allow the Inspectorate to define the remaining hold points.

To a large extent, for new plant, the Inspectorate base their selection of what to examine and agree to on the preliminary design proposals. Generally, for storage facilities the stages involved up to operation include;

- the start of construction

- the start of commissioning

- the introduction of radioactivity/active commissioning, and

- the commencement of operations.

The Inspectorate's agreement allowing progress is subject to the submission of a safety case by the licensee. This has to be found to be acceptable after assessment by the Inspectorate. The NII also uses the results of inspection to confirm that adequate arrangements are in place to control construction or commissioning activities and that these are being implemented satisfactorily so that the plant meets the specifications, limits and conditions set out in the safety case.

On operating plant annual reviews of safety by licensees are used to confirm that future operations will be in accordance with the operational safety case. These reviews are also used to ensure progress on any significant safety issues which may have arisen from the Inspectorate's assessment or inspections. The licensees can use these reviews to provide information for the periodic review of safety. Again the safety case produced as a result of periodic safety review informs the Inspectorate of which aspects of the licensee's plant or arrangements should be inspected. Inspections also inform and guide the Inspectorate's assessment of the safety case.

1.4   Safety Assessment

NII has its own 'Safety Assessment Principles for Nuclear Installations[1]' (SAPs) developed primarily for use by its own staff, but also with a view to assisting designers and operators. These principles form a statement of NII's views on particular aspects of design safety assessment related to future and existing plant. Of the 333 SAP's, 21 deal specifically with radioactive waste or scrap materials (residues). These are complimentary to and support NII policies. The policies and principles applying to radioactive wastes are considered below.

2. WASTE MANAGEMENT POLICY AND REQUIREMENTS FOR WASTE STORAGE

2.1 NII Responsibilities in Regulation of Radioactive Waste Management

Overall responsibility for national policy on the disposal of radioactive waste rests with the Department of the Environment, while the management of radioactive waste on nuclear licensed sites is regulated by the Health and Safety Executive's (HSE) NII. It is NII policy that radioactive waste is managed in accordance with the national policy, to standards acceptable to NII and in accordance with the disposal related interests of the Environment Agencies. Consultation through memoranda of understanding is used to ensure that NII imposes and then enforces licence conditions which require the licensee to manage on-site radioactive waste in a manner acceptable to HSE and the Environment Agencies. Arising from national policy as expressed in the recent Government "white paper" that reviewed radioactive waste management policy[2] and a consideration of our functions, NII have developed a range of position statements. These are supported by the NII's Safety Assessment Principles. (Extracts from the "white paper" are shown in italics in the remainder of this section and in cases are embodied directly into NII policy which is shown in bold type).

The underlying principle of national policy, NII position statement and Safety Assessment Principles is that waste should be safely and appropriately treated, stored and managed. This applies equally to all radioactive materials whether waste or not.

Many of the position statements described below are equally relevant to plutonium storage and other materials which are not declared to be wastes. Section 3 discusses some of the safety issues associated with plutonium which place a particular emphasis on the aptness of particular policies.

2.2 Generation and management of waste

The "white paper " on waste management states:.....*the Government will maintain and continue to develop a policy and regulatory framework which ensure that:*

(a) radioactive wastes are not unnecessarily created;

(b) such wastes are safely and appropriately managed and treated;

(c) they are then safely disposed of at appropriate times and in appropriate ways;

The unnecessary generation of wastes recognises that costs and dose have to be taken into account and, for example, minimisation of volume by compaction or concentration may not in all circumstances be desirable.

2.3 Strategic Planning of Radioactive Waste Management

The Government's policy is that: "....*producers and owners of radioactive waste are responsible for developing their own waste management strategies consulting the Government regulatory bodies and disposal organisations as appropriate. They should ensure that:*

a)   they do not create waste management problems which cannot be resolved using current techniques or techniques which could be derived from current lines of development;

b)   where it is practical and cost-effective to do so, they characterise and segregate waste on the basis of physical and chemical properties and store it in accordance with the principles of passive safety (i.e. the waste is immobilised and the need for maintenance, monitoring or other human intervention is minimised) in order to facilitate safe management and disposal;

c)   they undertake strategic planning including the development of programmes for the disposal of waste, including the development of programmes for the disposal of waste accumulated at nuclear sites within an appropriate timescale and for the decommissioning of redundant plant and facilities. These programmes should be acceptable to the regulators and discussed with them well in advance.

NII requires licensees to undertake strategic planning for radioactive waste management, including the development of programmes for the disposal of waste accumulated at nuclear sites within an appropriate timescale.

NII expects that radioactive waste management should take an appropriately balanced account of the radiological risks to workers and to the public including potential doses from accidents.

NII's Safety Assessment Principle P295 also reflects this approach. It states that: "The generation of waste of a type or form incompatible with currently available storage or disposal technology should be avoided."

2.4 Site Specific Waste Strategies

NII requires licensees to develop a site specific strategy which provides for the management of all radioactive waste on site with particular emphasis on the long term safe management of waste for which there is no authorised disposal route.

2.5 Continuity of radioactive waste management responsibilities throughout a licensee's period of responsibility

NII will continue to take appropriate regulatory action where necessary to ensure that licensees undertake the on-site management of radioactive waste to standards that are

acceptable to NII throughout the period of the licence and any subsequent period of responsibility.

2.6   Segregation and characterisation of wastes

NII requires that, where it is practical and cost effective to do so, radioactive waste should be segregated and characterised in order to facilitate safe waste management and disposal. This follows from the statement expressed in section 2.3 (b) above.

2.7   Disposal of radioactive waste

Wastes should be safely disposed of at appropriate times and in appropriate ways.

2.8   Passive safety features

When it is necessary to store radioactive waste, NII requires that, where it is reasonably practicable to do so, it is stored in a passively safe form and in a manner which facilitates retrieval for final disposal (see also section 2.3(b) above).

2.9   Retrieval or transfer of stored waste

NII expects that any future waste storage facilitates or modifications to existing facilities should be designed to facilitate retrieval and transfer of waste.

Safety Assessment Principle P296 states that "Radioactive waste stored on site should be in a form which minimises the hazard of storage and is compatible with retrieval and with any subsequent storage, transport or disposal route, and it should be appropriately monitored and inspected to ensure that it remains in such a form."

This principle applies equally to all radioactive materials. Any irretrievable materials would become waste by default and for which methods of retrieval would need to be developed.

2.10   Projected use of storage facilities

NII requires that waste storage facilities should:

-    have a valid safety case for a specific design life or projected life;

-    have an appropriate maintenance and surveillance programme; and

-    have a full periodic safety review against modern standards to substantiate the projected life.

Appropriate methods to restrict doses should be employed when it is necessary to retrieve and transfer stored waste from such facilities.

This position statement is supported by site licence conditions which require safety cases to be periodically reviewed and the examination, inspection, maintenance and testing of plant or related items. In addition, SAP P298 states that:-

"Adequate provision should be made for:

(a)   monitoring and maintaining in a safe state accumulations of stored radioactive waste;

(b)   determining and recording appropriate details (e.g. quantity, type, origin and form) of the radioactive waste or scrap in a manner which is durable for the anticipated period prior to final disposal;

(c)   estimating the rate of arising and transfer, the change of volume on conditioning and the volume and activity of the waste or scrap in each store."

Properties that will typically be monitored for raw liquid wastes include volume, temperature, pH, solids content, etc. Monitoring of solid wastes is likely to be less extensive, although for example humidity and temperature of the storage environment may be measured. Necessary inspections must be carried out as required by the site licence. Visual inspection of packages may give early warning of degradation of packaging materials and dimensional changes.

2.11.   International Standards and Developments

Radioactive waste management should be undertaken to internationally acceptable standards. Where standards or guidance produced by international consensus such as those of IAEA exist, NII will take these into account in assessing the acceptability of radioactive waste management safety cases. NII will maintain awareness of, and involvement in, national and international developments in the field of radioactive waste management.

3.   **PLUTONIUM STORAGE REGULATION**

3.1   The NII's approach to Plutonium

There is now a substantial stockpile of plutonium in the world, arising from the reprocessing of used nuclear fuels and the decommissioning of nuclear weapons. While consideration is being given to the beneficial use of this material in mixed-oxide fuel, the hazard to the public and workers of storing plutonium will remain for some time. Vigilance continues to be necessary to ensure that the risks are properly controlled.

NII's approach to the regulation of plutonium is the same as for other materials and activities associated with the nuclear industry. However, NII has recently developed its thinking and approach to waste management to a greater extent. Many of the resulting waste management position statements are applicable to plutonium management and, also the management of other non-wastes.

3.2   Plutonium Safety Issues

There are some difficult technical, safety and strategic challenges associated with plutonium and plutonium residues. What are these and how do they fit in with requirements for waste management?

Packaging

The fitness for purpose of packaging must be kept under continuous review, particularly as the timescales for re-use or disposal may extend well into the future for much of the material. There are, however, potentially more immediate problems.

Some plutonium residues in the UK have been stored in non-standard packaging. These may be unsuitable for continued use since it was originally planned to store them for only a short time pending recovery. Repackaging and reprocessing may eventually be required in some cases and to do so will generate additional waste and dose-uptake by personnel. Careful judgements must be made. The benefits of safer storage, must be balanced against the short term risks associated with repackaging etc. Factors such as the likely period of storage pending re-use or disposal must be taken into account in determining the acceptability of the chosen strategy. Established waste management policies are adopted and applied to repackaging to cover:-

- Segregation and characterisation of radioactive materials (see Section 2.6).

- Passive safety features (see Section 2.8).

- Retrieval or transfer of stored materials (see Section 2.9).

- Projected use of storage facilities (see Section 2.10).

Processing and Recovery of Residue

Plutonium produced by current technology is in the form of plutonium dioxide which can be stored under passively safe conditions. This is not necessarily so for historic residues. Some plutonium residues may be pyrophoric and unstable. Others can absorb moisture from ambient air, becoming more corrosive, swelling or releasing flammable gases. Long term storage of these materials is undesirable and they are candidates for reprocessing either to recover plutonium into a standard form or to convert it to a passively safe and stable form suitable for long term storage as waste.

Again, existing waste management policies are adapted and applied to cover:-

- Segregation and characterisation of radioactive materials (see Section 2.6).

- Passive safety features (see Section 2.8).

- Retrieval or transfer of stored materials (see Section 2.9).

- Projected use of storage facilities (see Section 2.10).

Plutonium Risk Management

With plutonium dioxide and plutonium residues, the material in storage is gradually changing as americium is produced by plutonium decay. Radiogenic heat and radiation hazards increase with time, making further processing and handling difficult. This is another factor which needs to be taken into account when reviewing safety. It requires a consideration of the likelihood of the long term degradation of safety and the need to intervene by, for example, reprocessing or repacking sooner rather than later in order to avoid dose uptake penalties. This is a requirement not only to take a forward look at needs and potential problems but also a requirement for a proactive risk management strategy. The periodic review of safety, which is a licence condition requirement, provides a focal point for this work.

Again established waste management policies are adapted and applied to cover:

- Site specific strategies (see Section 2.3).

- Segregation and characterisation of radioactive materials (see Section 2.6).

- Passive Safety (see Section 2.8).

- Retrieval or transfer of stored materials (see Section 2.9).

- Projected use of storage facilities (see Section 2.10).

3.3 Safety Assessment of Plutonium Stores

It has been shown that most of the policies and safety assessment principles which have been developed for radioactive waste management on licensed sites are equally relevant to the long term safe storage and management of nuclear matter in general. In particular, the same standards of safety apply to the storage of plutonium and plutonium residues, as to radioactive wastes.

4. CONCLUSIONS

UK law places the responsibility for safety with the licensees. It establishes a goal-setting framework which is flexible and allows licensees to prepare arrangements for the management of safety which suit their organisational structures. These arrangements must provide adequate opportunities for the regulator to select and examine how safety is being managed.

Licensees must also have regard to national policy and the requirements of the safety regulator and Environment Agencies when developing proposals for waste storage.

The requirements of the regulator for the storage of radioactive waste are based on experience, reinforce national policy and represent good practice for the safe storage and strategic management of these materials.

The Government "white paper"[2] reaffirms that some potentially useful radioactive materials (such as spent fuel, reprocessed uranium or plutonium) are not regarded as wastes whilst further use remains an option. Nevertheless, the NII position statements, and safety assessment principles for radioactive wastes which are discussed above are equally relevant to the long term safe storage and strategic management of plutonium and these other materials.

5.   REFERENCES

1.   "Safety Assessment Principles for Nuclear Plants", HMSO 1992.

2.   "Review of Radioactive Waste Management Policy", Cm2919, HMSO 1995.

"The views expressed in this paper are those of the authors and do not necessarily represent those of the Health and Safety Executive."